UNCHAGAH[1]

*Life by the Upper Peace River
before the Dams, 1928 – 1932*

by HENRY STOTT

Suite 300 - 990 Fort St
Victoria, BC, V8V 3K2
Canada

www.friesenpress.com

Copyright © 2017 by Charles Stott
First Edition — 2017

All rights reserved.

No part of this publication may be reproduced in any form, or by any means, electronic or mechanical, including photocopying, recording, or any information browsing, storage, or retrieval system, without permission in writing from FriesenPress.

ISBN
978-1-4602-9124-5 (Hardcover)
978-1-4602-9125-2 (Paperback)
978-1-4602-9126-9 (eBook)

1. Biography & Autobiography, Personal Memoirs

Distributed to the trade by The Ingram Book Company

"A man should walk on sand before he sleeps in silk."

—Arabian Proverb

To

T.S.S.

*"Pervixi: neque enim fortuna malignior unquam
Eripiet nobis quod prior hora dedit."*

(I survived: For luck was never crueler.
The next hour will steal from us what
the previous hour gave)

—Petronius Arbiter

Acknowledgement

HAVING READ THE account of my uncle Henry's five years spent with my father living by the Upper Peace River, I decided to visit the Hudson's Hope area for the first time during the late summer of 2005. I took with me copies of many old photographs from father's collection.

In Hudson's Hope I was able to meet with several members of the Beattie family and Ethel Rutledge, several of whom had met my uncle and father at the time of the account. Without their assistance we would have been unable to identify many of the people and places shown in the photographs. Their knowledge of the area proved invaluable and we are much indebted to them for sharing this with us.

While helping my cousins to prepare the manuscript for publication I have also received useful information and assistance from various archives, museums and universities which has also been much appreciated.

TED STOTT,
April, 2017

Contents

Preface I

The Quest 1

The Journey 12

Peace River 37

The Return 107

Appendices 135

Index of Images 150

Henry Stott, 1947

Preface

I WAS BORN in February and during the following summer I was taken on fine days in a perambulator and parked under a magnificent copper beech tree. There I lay, gazing up at the entrancing greenish-brown leaves, conscious from time to time, of objects, some larger, some smaller, flitting across my field of vision; warm, comfortable, and helpless.

Nowadays, in summer, I wake at about six o'clock in the morning, draw my bedroom curtains and, sitting up in bed, look out over the fields and hedgerows to the trees beyond. There is a constant coming and going outside my window, and this I know to be the house martins flying to and from their nests under the eaves.

I have learned to focus my eyes, to give things a name and to note sequences of events, and I have been able to use this knowledge to advantage. But, although in the course of my life I have done all the usual things and some not so usual, my feelings on these early morning occasions are much the same as they were some seventy-five years ago; feelings of warmth, comfort, a slight consciousness of the latent hostility in the nature of things, and a vague wonder as to what it is all about.

If man had been able to view the solar system from outside, he would not have taken so long to discover the nature of its geometry, and at times one feels that a glimpse of life from an unusual angle may be similarly revealing. This is an account of one of my less common experiences. Undertaken in a spirit of expectation, it proved to be one of the most enjoyable in my life. But it led nowhere, and in the end I returned to the kind of life into which I had been born, satisfied that there was no way back and no way out. Not that there is much satisfaction in watching at close range what seems to be the decay of a civilization; but an ethos cannot be produced out of a top hat, much less out of a cloth cap. One day, no doubt, the phoenix will return to Heliopolis.

H.S.
Brant Broughton
February 1970

The Quest

"We are never in ourselves, but beyond."

—Montaigne

I WAS HUNGRY and tired, and darkness was already falling as I climbed the last hill. Dirty, too, for I had not washed for several days. I hoped that, this time, my brother Tom[2] would have arrived before me; then the cabin would be warm, ice chopped and melted, and a meal, perhaps, already prepared. I paused a moment at the top of the hill and looked down into the valley below where the Peace River, running now under four or five feet of ice, swept in a great curve eastward from the heart of the Rocky Mountains. Complete silence and an intense cold enveloped the land, but a shaft of light from the window told me that the cabin was occupied, and the prospect of warmth and food cheered me on the last half mile.

I reached the cabin, slipped off my snowshoes and pack, and leant axe and rifle against the veranda wall. Then I opened the door. Tom had arrived some time ago, bathed, shaved, and dined, and was now lying on his bed in dressing gown and yellow silk pyjamas with a large cigar in his mouth and whisky on the floor beside him. A sight, I thought, probably unique in all the trapping cabins in the north of Canada. Dirty and unshaven, I leaned against the door jamb and laughed, letting in a stream of freezing air. Tom saw the open door, but not the joke; he shivered and cursed.

I entered and shut the door, reflecting, ironically, on the change that had taken place in our attitude to physical hardship. For two years we had slept on spruce boughs and used a mackinaw coat as a pillow. We had welcomed hardship as a tonic, and slowly the tonic had done its work. I thought of old Louis Petersen in his cabin at Finlay Forks, eating bully beef and potatoes, and remarking, "Nothing's too good for me if I can get it." We had come to see life through the eyes of Petersen and his kind.

The result was curious. The lamp light gleamed on the polished oak floor of the cabin. There was a radio, whisky, cigars, and comfortable beds. But there were no chairs. We had not felt the need of chairs and still sat on soap boxes whenever we wanted to sit down. There were no pictures, either; guns, revolvers and a shelf or two of books were the only decorations on the walls. No curtains covered the windows, and there were no sheets on the beds. We had everything we desired and the almost equal satisfaction of having nothing that we did not want.

"There are two things in life that can be wrong with any man," says Sir Patrick Cullen in "The Doctor's Dilemma", "one is a cheque. The other is a woman". It may be we owed our carefree existence to our failure to find gold in payable quantities, or to meet any of those attractive damsels in distress with which fiction of the North American continent abounds. Certainly we had little to do with the root of all evil—whichever of the two it may be. Once a year we visited civilization, sold fur and bought supplies. Once a year we gazed on the civilized female toilette with a detachment that we might have found difficult to preserve if we had looked too long. Then we came back into the bush.

We had achieved this satisfactory life almost as casually as one goes away for a weekend. Early in 1928 my brother Tom, who spent some years chicken farming in southern British Columbia, paid a visit to England. He had tired of his settled life, and he found me in a similar mood, although for different reasons. Ever since Mussolini had been allowed to get away with the shelling of Corfu in 1923, and Mr. Vernon Bartlett had commented that, after all, it was not the business of the League of Nations to secure justice but to preserve peace, I had felt that another war was only a matter of time. And I had not yet learned the philosophy of building on the slopes of a volcano which, now that the war has come and gone, is our common heritage. I was quite ready to go while the going was good.

Time, I felt was limited, and there was a question to answer—a question posed by the extreme complexity of civilization and the discipline that it imposed, contrasted with the brutal and primitive simplicity of what appeared to be its ends. Was it worth while?

When one starts to ask that question there are two roads one can travel. One can immerse oneself in music, literature, art, drink or drugs and live in a world of fantasy, or one can place oneself in a position in which one has to take energetic and possibly unpleasant action if one wishes to live at all. If you have to kill your deer, gralloch[3] it, and butcher it before you can cook a steak, you will soon find out where you stand. If you are not prepared to be overwhelmed by life, you must assume that you believe in life, whatever depths of anger or despair it may provoke. You may not find the reason for it, but you may find the courage to live it.

It was now three years since these thoughts passed through my mind, and, as I pondered on those years, they seemed almost too good to be true. So good that my comfort was disturbed by a vague disquiet. It is not lucky to be too conscious of good fortune. I broke the train of my thoughts, half fearing that something might go wrong with my world.

———

There is almost everything in Canada that a man could want, but to me it is pre-eminently a country of rivers and lakes. These are well known for their beauty and the sport they afford the fisherman. It is less often remembered today that for many thousands of miles they provide the most delightful travelling in the world - a transport system marvellous both in its nature and in its extent. One can travel from Lake Winnipeg by river and lake into the Atlantic Ocean, the Pacific, the Arctic and the Gulf of Mexico. These endless miles of waterways are broken by unnavigable rapids and canyons and by an occasional portage over the land between one river and another, but none of these portages need take longer than a single day.

One can travel for days without seeing any trace of man, but the rivers are alive with Canadian history. They were the routes of the old explorers who opened up Canada, and, in the west, penetrated as far south as the mouth of the Columbia River. It was over them that a hundred and fifty years ago the great battle for the fur trade of Canada was fought between the Hudson's Bay Company with its headquarters in London and the North West Company of Montreal. As the place names attest, it was largely a battle between Scots. Scotland, the Islands and, in particular the Orkneys were a great recruiting ground for the Hudson's Bay Company. The names of the partners in the North West Company—the McDonalds, McGillivrays, Mackenzies, and many more—speak for themselves. The battle nearly ruined both companies and ended after many vicissitudes in a victory for the Hudson's Bay, and an amalgamation in 1821.

These rivers remained the highways of Canada for many years, but at the time of which I write they had, for the most part, seen their day of service. There was one exception—the Mackenzie River system. Here only an occasional aeroplane fitted with pontoons in summer and skis in winter, and the outboard motor bore witness to the passage of time. Winter and summer the rivers and lakes were still the highways of the Canadian north-west, and the headquarters of trapper and prospector, and the Hudson's Bay Company's posts were to be found on their banks and shores. The

Mackenzie River and its tributaries form only a part of a much larger system, but from Summit Lake, its farthest point south, to Aklavic in the Arctic Circle, is a continuous waterway of two thousand five hundred miles.

One of the most remarkable links in this chain of rivers and lakes was the Peace River. It is formed some two hundred miles north of Summit Lake on the western slopes of the Rockies by the junction of the Finlay and Parsnip rivers at a point known as Finlay Forks. Thereafter it runs due east, cutting through the entire mountain range, a distance of about one hundred miles. Eighty miles is navigable water, broken by the Finlay and the Ne Parle Pas rapids. Then follows the Rocky Mountain Canyon, twenty-two miles long, and then the prairie country known as the Peace River Block[4]. In cutting its way through the Rockies the Peace is almost unique. No other river north of the Rio Grande performs this feat, except the Liard.

One hundred and sixty years ago Alexander Mackenzie passed up the Peace River with a crew of Iroquois Indians on his way to the Pacific coast. It was then almost a virgin land. Eighty years later, when Captain Butler[5], as he then was, visited the country, it was unchanged save for the presence of an occasional white trapper and one incredible black man who was waging a one-man war against the Company's post at Hudson's Hope. The overflow from the Cariboo and Manson Creek gold strikes added a few prospectors, and later, during the gold rush to the Yukon and Alaska, a number of unfortunates tried the overland route from Edmonton via the Peace River and the Laurier Pass, and most of them left their bones by the way.

Alluvial gold was found in many places on the river bars, but never in any great quantity, and there was, it seemed, no substantial body of ore. So, despite Captain Butler's advocacy of a railroad through the Peace Pass, the country remained undeveloped, inhabited only by a few trappers, some of whom prospected during the summer months when fur was out of season.

Forty years ago the outboard motor had taken the place of the spruce pole on the river, and horses were not unknown. Indians were fewer, white trappers more numerous, and the white man had formed the habit of building small cabins on his trap line instead of "Siwashing-it" all winter with a piece of canvas and half a Hudson's Bay blanket. Keen frosts and high water changed the runs and eddies of the river from year to year, and forest fires marred the spruce on its banks. Little else had changed. The Upper Peace River[6] was still primitive, remote, and quiet.

In that restless period between the wars, the attraction of these magnificent waterways lured us, and visions of changing vistas round countless river bends and of innumerable trout lurking in the depths, proved irresistible. There seemed a promise, too, in life in this country—the promise of freedom and a quiet mind, of sanity and stability—which civilization lacked forty years ago, and lacks even more today. We acted very much on the spur of the moment and without any very definite plans. We knew that we could live largely off the country and earn by trapping what little money we might need. We had no ties and no time limit, and were free to take an extended holiday or discover a new way of life.

So we left England and travelled west.

SS Laurentic, June 1, 1928 – Tom Stott on left, age twenty-four, during the trip from England to Canada and the Peace River with Henry Stott.

As I take little interest in sight-seeing, I have crossed Canada several times without ever stopping to look at Niagara Falls. I am bored by being carried about in ships, trains, and aeroplanes, and Tom is even less interested. On this occasion we varied the experience by travelling as immigrants, although Tom had already lived in Canada several years and my movements were uncertain. Vaccination and the possession of one hundred dollars were the conditions of our entry; in fact, we took an extra two hundred dollars as we did not expect to earn anything for a least twelve months.

From their speech I judged that most of our fellow immigrants came from Glasgow. Scotland was still contributing its quota to populate Canada. I was happy to find several people on board who played chess, including a Russian who said that he had played against Alekhine[7]. This might well have been; he was certainly the best player on board. But he was addicted to the bottle, a habit which, he said, he had acquired while crossing the Pacific on a sailing ship, and I found that if I could make the game last long enough, I could beat him. Unlike most Russians he did not insist that I drank glass for glass.

We spent about two hours in Quebec and then boarded an immigrant train on which we lived for the next few days. It was not a comfortable journey, and if there is anything more boring than crossing the Atlantic by ship, it is crossing prairie country by train. Oceans and prairies have one common attraction—magnificent sunsets. I have seen many of them on the plains of eastern Poland (now western Russia), and I shall always remember watching the sun set down the fjord from a hill near Bergen. But sunsets are not everything, and there are twenty-four hours in a day.

There was only one incident during the journey. We were sitting on opposite sides of the carriage, leaning towards each other in conversation, when there was a sharp crack, and a hole made, apparently, by a .22 long rifle bullet appeared in the centre of the window beside us. Someone had taken a shot at the train. The conductor did not seem to think this a matter of importance. Perhaps he did not like immigrants.

After crossing the prairies, we broke the journey for a few hours at Edmonton, the capital of Alberta. In those days the prosperity of Edmonton depended chiefly upon wheat. The Turner Valley oil field was being opened up, but the great strikes were still to come. If we had gone to Canada ten years later, we might have made money in oil. Apart from the strikes near Edmonton, natural gas is now being piped from the Peace River Block, and I have often seen oil seepage in the mountains to the west. But we were too early, and we lost what little money we invested in oil.

Leaving Edmonton, we crossed the Rockies, passing through magnificent mountain scenery, and finally reached the town of Prince George on the Pacific side of the range. Prince George lies near the northern bend of the Fraser River, about four hundred miles north of Vancouver. The hills that turn the Fraser from its north-westerly course form the watershed that divides the rivers running into the Pacific from those running into the Arctic Ocean, and thirty-two miles of road have replaced the old Giscome portage that connected the two waterways. Summit Lake, at the northern end of this road, was the last outpost of civilization.

We found Prince George to be a town of about four thousand inhabitants. It was built almost entirely of wood. Log houses, reminiscent of the days when it was Fort George, still stood on the bank of the Nechako River, but life was no longer so hectic as in the days of the building of the Grand Trunk Pacific Railway[8]. Prince George had settled down as a centre of the logging industry.

It was also a centre for the trapper, and every year early in May, as soon as the ice had left the river and lake, trappers from the Finlay, the Parsnip, and the upper Peace rivers made their way several hundred miles upstream to Summit Lake, and thence by road to Prince George. Here they took their annual relaxation, and outfitted themselves for another season. They seldom stayed in town long. Many spent the summer months prospecting, setting out as soon as possible in order to take advantage of high water in the rivers. We arrived in Prince George on the sixth of June and found it already empty of these seasonal visitors.

It was hot and dusty, and we had no desire to linger there; our own departure was overdue. But an outfit and supplies for an exploratory trip of three months had to be bought, and we could use all the information we could get about the north country. Our shopping was easy, and the cool, gloomy, general store a pleasant relief from the heat outside. We ordered provisions from a grocer across the street from the hotel. The store contained all else we needed: clothing, boots, moccasins, moose hide, tents, rifles, and hardware of all kinds. We shopped leisurely, sitting on boxes and bales and discussing conditions in the north with the proprietor and his customers. But their knowledge was all hearsay and little greater than our own.

We did meet one man who had made a trip into the Peace River country. Oddly enough, he was a taxi driver, but so, for that matter, was the man[9] who had formerly driven the stagecoach from Prince George to Vancouver. Our taxi driver was an ex-cowboy. Until 1913, when the railway was built, Prince George had been an old-fashioned frontier town, and some of its inhabitants belonged rather to the old time than the new. They had been overtaken by an advancing civilization and had lacked either the will or the desire to escape.

This taxi driver had been a passenger on the river trip and could give us little information of value. He was far more interested in the merits and demerits of various local Indian tribes than in the rapids of the Peace River. But he had for sale a heavy aluminum camp cooking outfit and a .22 Remington rifle. A friend of his, who had frozen his capital so completely in a string of packhorses that he lacked ready money to buy a shirt, offered us a 30.30 Winchester carbine. The Winchester is the trapper's usual weapon. It carries a heavy bullet and its short range is adequate for a man who does most of his shooting at two hundred yards or less.

We spent an afternoon bargaining over these articles, all of which we bought. Between whiles the man with the packhorses and the taxi driver discussed Indians and the folly of educating them and extending the protection of the law to them. "The

Chilacotans[10] sure has you fellers bluffed" said the taxi driver to a Siwash[11] Indian who squatted gloomily in a corner. "A Chilacotan's worth a whole camp of Siwash any day." "Them damn bad Indians" agreed the Siwash, morosely.

It seemed that a Chilacotan could also bluff a white man, for the taxi driver then told us, as an example of the acquired iniquity of the Indian, how an "educated" Chilacotan chief named Jerome had once driven a string of horses through the middle of a settler's grain field (it was an old Indian trail) and when the settler came out with a gun and threatened to shoot him, had remarked pleasantly "You no shoot. If you shoot maybe you hang." The taxi driver gave credit where credit was due, concluding judicially, "A goddamn fine Indian, educated by a Goddamn fine crooked, white man." His friend with the packhorses was more practical. "I'd sure have taken a chance at them horses," he remarked after some thought. "There ain't no death penalty for shooting horses, anyway." The Siwash said nothing; his gloom was unrelieved, but one wondered what he was thinking.

There is a story of a Salvation Army meeting at which an Indian convert was taking up a collection. An old Indian standing by was observed to be deeply amused, and, in answer to a question, remarked with relish: "Long time ago white man fool Indian; now Indian fool white man."

We kept leading the conversation back to the Peace River, but only one piece of information rewarded us. The Arctic trout in the Wicked River—a tributary of the Peace—were, we were told, incomparable, and the place to catch them was in a pool about three and a half miles upstream where a natural bridge of rock spanned the river. We instantly acquired one definite, if minor, objective—to catch and eat these trout. It was to remain an ambition for a long time. We had been in the country three years before we achieved it.

The next day we completed the purchase of our outfit. It was simple enough. Khaki shirts; khaki drill trousers; heavy woollen underwear and socks; shoepacks; moccasins; Stetson hats; one four-point Hudson's Bay blanket; an 8' x 6' tent fitted with mosquito netting and ground sheet, and a tarpaulin to cover our supplies in the boat; knives; an axe; shovels; gold pans; pick and a sand screen; a roll of snare wire; fifty feet of manila rope; ammunition for the rifles; fishing tackle; and cooking utensils. These were bare necessities, and, apart from a minimum of staple foods, these were all we took with us. We hired our friend with the taxi to take us out to Summit Lake.

The thin ribbon of road winding north from Prince George runs through hilly country thickly covered with spruce and poplar. There are firs, too at Summit Lake, although it is near their northern limit. These are not comparable with the giant Douglas fir of the coast, but they are sizeable trees, up to three feet through at the butt. Most of the timber is spruce, and it grows down to the water's edge both on the mainland and on the many small islands which dot the lake.

Here one begins to realize how much the pleasant diversity of the English landscape is due to the variety of its trees. For hundreds of miles north this country is heavily wooded, but here, too, is something of the austere spirit of desert and prairie. Among the countless miles of spruce, interspersed with occasional patches of poplar, there are other trees to be found; balsam or cotton wood, from which dugout canoes were formerly made, and which is said to have some healing virtue in its buds; a very few birch, the only hard wood; small willow in swampy ground; lodge pole pine; and a few yellow pine. But these need looking for; they do not spring to the eye and vary the contours of the forests which are severe, like the climate, and less kindly to the eye than English woodlands.

At the south-eastern end of the lake there was a small store, the Hudson's Bay company's warehouse, (it was from here that the company freighted in supplies to its posts at Fort Grahame and Whitewater on the Finlay River), a few log cabins, and the boat builder's shed. The wooden boat had long ago replaced the canoe, birch bark or dugout, although there was one purpose for which a light canoe was still used. It provided the only means for crossing a river just before freeze-up, when there may be eighty to a hundred feet of shore ice, and slush ice running in the current. Then a light canoe can be carried by one man across the ice, launched, paddled through the drift, and dragged out over the ice on the other side.

Shallow draft was essential, and for general purposes a flat-bottomed boat, about forty feet long with a beam of six to seven feet was used. A boat of this size will carry about three (short) tons of freight. Paddle, oars, and the spruce pole, in use until the early nineteen twenties, had now been supplanted by the outboard motor. If one could run in from two to three feet of water and take every advantage that the current had to offer, the old Johnson 8 hp outboard fitted with a three bladed 12" x 12" propeller would push a fully loaded boat upstream. The course steered is a zigzag, following one bank of the river until further progress is impossible, then drifting across and clinging to the opposite bank, and one's speed averaged less than four miles an hour; but it was easier and quicker than the spruce pole.

We were to discover in time why the forty-footer was in general use. It was as large a boat as the outboard could handle. It was also near the maximum length to take the bends in the Crooked River, which was unnavigable for a boat fitted with an inboard engine except in extremely high water. Thirty-five feet was about the minimum length for running the Finlay and Ne Parle Pas rapids with safety, and that left only five feet to argue about. But for the argumentative there was still plenty of scope for discussion in the flare of the sides and the amount of rake fore and aft.

Our ignorance at this time was not of any great importance. We were going downstream under our own power, and, as we had numerous lakes and stretches of dead water to cross with possible head winds, we purchased the smallest boat we could get. This boat had, in fact, been designed only for use on the lake. It was twenty-seven and a half feet long—big enough to hold us and our outfit comfortably. We decided to line down the rapids if we did not like the look of them.

We transported our outfit and supplies to the Lake on the thirteenth of June, and, after fitting a handle to the axe, loaded up and left early in the afternoon. Paddling across the Lake, keeping Teapot Mountain on our left as we had been directed, we found the outlet without difficulty, and glided slowly through the reeds and lily-pads into the peaty-brown waters of the Crooked River. It is over sixty miles from Summit Lake to Lake McLeod, and, like the mile in the nursery rhyme, most of these are crooked. The first twenty miles are also very swift with numerous riffles (the diminutive of "rapids"); successions of swift chutes with many "S" bends and deep pools. The most notable of these, known as the "Long Riffle" was upwards of a mile and a half in length. At many of the "S" bends cribbing had been put in to prevent the water dissipating itself in the surrounding muskeg. These unexpected signs of man's handiwork were the result of some expenditure of government money; the rivers, after all, were the roads of the north.

We camped that evening at about six o'clock, or rather, we landed with the intention of making camp. But Tom was so glad to be out of civilization that it was hours before he could bring his mind to consider putting up the tent and preparing food. So we sat and smoked, and slowly a soft rain began to fall on us. We needed time to adjust to the changed momentum of our lives. Up to now we had travelled quickly; it was less than three weeks since we had left Liverpool, and our impressions were chiefly of noise, heat, and dust. Now our speed had slowed down from many thousands of horsepower to a two manpower standard. Our control over nature had decreased enormously. There was little we could do but travel and sustain life, and that with some effort. But, at the moment there was little else we wanted to do, and if nature had become a harder taskmaster, human society had relaxed its grip, and our personal freedom had become almost unlimited.

As the rain slowly soaked into us, we adjusted our minds to these changes with a feeling of deep contentment. It is easy and delightful to throw off the shackles of civilization. Its benefits are dubious, its defects obvious, and it is all comparatively recent history. Man's servitude to the slow routine is very old, and to return to it is almost like coming home. This feeling is very powerful, but is it only a mood, or can it, as some have thought, form the basis of a philosophy of life? Perhaps it is only the natural inertia of the human spirit that makes one feel at times like these that the price of civilization is too high.

For me this was an intriguing question, and one which I knew would take some years to answer. No question existed for my brother. He had endured the journey to Prince George with patience and, for the most part, in silence. Now he was where he belonged, and there was even less need to talk.

We dreamed happily for a while. Then I came out of my reverie, for I was not yet conditioned to sleeping in damp underwear. I roused Tom. We put the tent up, and supped off tea and rye bread. Not much of a meal, but it suited our mood. It contained one other ingredient—a dash of the cleaning powder we had used in the billycan. This gave a queer flavour to the tea and, to the credit of the preparation, some of the tartar came off my teeth.

Then we cut some small spruce boughs and covered the ground underneath the tent with them, spread our ground sheet, and rolled up in our respective halves of the Hudson's Bay blanket. Sleep came slowly the first night under canvas, but we had left behind us the noise of steamship and railroad, and it was almost as restful as sleep itself to lie and listen to the murmur of the river and the occasional creak of a rope as our boat swung idly in the eddy where we had moored it.

The Journey

"Gleams that untravell'd world, whose margin fades
For ever and for ever when I move.
How dull it is to pause and make an end..."

—Tennyson

RUNNING WATER HAS two outstanding qualities—a continuity which is almost hypnotic, and a variety which has infinite charm. These formed a continuous background to our lives for the next five weeks and were never far away during the next five years. The never-ending flow of the river carried with it a sense of destiny in keeping with our expedition. Every mile we travelled downstream made it more difficult to return, and, although we could eventually find our way back to civilization overland to Edmonton, we had first to cover at least three hundred miles of river. We were committed, and we had the ease of mind which follows an irrevocable decision.

There are three ways in which I have enjoyed the changing movement of running water. One is wading up the bed of a river fishing a dry fly. This is an experience which I do not associate with Canada, and to my mind it is the best of all. The continuity of the river and the variety of its flow as a background to one's concentration on fishing produce a dream-like ecstasy which is among the most satisfactory experiences in life. The other two are paddling or poling a canoe, and navigating a rapid with an outboard motor. I was to become very familiar with these during the next few years, and both are very well worthwhile. There can be awkward moments. I have stood on my head in a river, waiting until my waders filled with water to right myself, and I have had some anxious moments in—and out of—boats. But to me running water is a friendly thing. I do not expect it to hurt me.

There was little danger or hardship in our lives as yet. If there had been, we should have welcomed it. Civilized man has to seek hardship as a tonic, and for some time to come we were unconcerned with comfort. This showed itself in the Spartan nature of our grubstake which we spent the following morning re-packing; for we had left Summit Lake hurriedly, and our supplies were all in bulk. We had estimated our needs for three months as follows:

100 lbs whole wheat flour; 20 lbs oatmeal; 20 lbs dried beans; 10 lbs salt pork; 10 lbs bacon; 2 lbs salt; 5 lbs tea; 5 lbs lard; 5 lbs chocolate; 4 lbs prunes; 4 lbs dried apricots; 5 lbs ship's biscuits 1 packet raisins; 1 packet yeast cakes; 1 lb baking powder; four dozen boxes of matches. A little tobacco was our only luxury. We took no jam, milk, butter, or tins of anything.

Our medicine chest consisted of a first aid outfit Sloan's liniment, ammoniated quinine, permanganate of potash crystals, aspirin, and some laxative pills of guaranteed violence. We also had some mosquito dope to rub on the skin, and found it effective for about ten minutes—long enough to get a smudge fire going. Later we came to depend on two cure-alls: castor oil and Hudson's Bay rum. The virtues of castor oil are well known; as for the rum it may be said that, in addition to being excellent rum, it was 150 proof spirit[12]. We suffered no ailment which could resist these two liquids, singly or in combination. Even toothache succumbed to the rum, although it left the gums with a white and shrivelled appearance that made on wonder what might be happening to one's liver.

For meat and fish, we depended on our own efforts, and we had not long to wait to fill our larder. We did not break camp until well after midday, and it was late in the afternoon when we shot the Harrison Riffle, half a mile or more of fast water ending in a long reach. As we came down the riffle with its succession of "S" bends, our attention was absorbed in keeping the boat from dashing itself to pieces against the bank, and our eyes were fixed on the eddies and swirls of the river. When the last bend was passed and all seemed plain sailing, we relaxed and glanced at the pool below.

Large hatches of flies are not uncommon, and occasionally one may see all the trout in a pool frantic with desire for them, but I have never seen anything quite so impressive as the foot of the Harrison riffle on that June afternoon. Travelling, I suppose at some ten miles an hour, our speed on that narrow stream seemed to us much quicker, and there below us in the pool was a regiment of trout. Side by side they were rising, nose, dorsal fin and tail. Where the rapid broke into the pool there scarcely seemed room for one more. The boat smashed into them. For a moment I had a feeling of impending massacre, and then, with scarcely a break, they were rising behind us.

There was a sandbar on our right and, without speaking, we swung the boat into it, stepped into the stream and, picking up our rods, which were lying ready for action, and started to fish. Tom looked calm enough. I felt that I must be trembling noticeably until, with the first fish on, the tension relaxed, and I became convinced that these trout were real.

As I fished I remembered a large hatch of iron blue dun which I had witnessed on the River Wharfe in Yorkshire some years before. Trout had risen furiously for three quarters of an hour, but I only succeeded in catching four. I was fishing dry, and my fly, as it came down the run into the pool, was to my eyes, and, I suspect, to the trout also, indistinguishable from the hundreds that surrounded it. There were so many flies on the water that, although the number of trout in the river was far greater than one had ever suspected, they could not take them all, and mine possessed no superior attractions to the natural insect. It held the mirror up to nature too faithfully.

On the Crooked River we were using Canadian flies; great birds, ten or twenty times the size of the flies the fish were taking. We were also using steel rods (I write it with a certain sense of shame, but no split cane or green hart could have survived with us for long). In those days, steel rods had a great deal of backlash, so that the fly made a nice splash when it fell on the water. The uneducated trout seemed to appreciate these efforts to attract them.

I do not know how long we fished. I was oblivious to all sense of time. I have no idea how many fish we caught for we freed them as fast as we caught them. I have used the word "trout", but the great majority were a coarse fish of a kind known locally as "squaw fish"[13]. These we threw back, keeping only four Rainbow trout weighing about a pound each for our supper. Then we made camp, and, although we had not come very far that day, decided that there was no reason to leave in the morning. The fishing had exercised its customary magic over our minds. We were under a powerful spell. As we lay in the tent looking out at the embers of our campfire, we thought no longer of the life we were leaving or of what lay before us, but fell asleep wondering how early in the morning the trout would be rising.

Fly-fishing is, without doubt, one of the most fascinating forms of human activity. One needs mental concentration and speed of action in many games, but seldom the sustained combination of both required to fish with a dry fly the runs of a clear, shallow stream. This, coupled with the delightful surroundings and the interest in selecting the right fly, take one completely out of the work-a day world. I never enjoyed fishing in its fullness in Canada. The trout were too innocent. But, unlike many fishermen of my acquaintance, I am fond of eating trout, so there were compensations. And never since those bountiful days have I been unduly interested in the size of my catch.

We slept soundly enough that night, and were up and fishing at half past five. It was a lovely morning and there were flies on the water, but the trout were sluggish.

Indeed, I have never had the luck to find trout rising well in the early morning as they seem to have done in the days of Izaak Walton[14]. We caught half a dozen in time for a seven o'clock breakfast, and after our meal, we tested the rifles and found the sighting accurate. Despite the time of the year a brace of grouse would have been welcome but none appeared, and Tom decided to vary our diet in a different way. We had no bread, and were eating bannock—flour, salt, baking powder, and water mixed into a dough and cooked in a frying pan. Just eatable when hot, but most unpalatable cold. Tom decided to bake bread.

In isolated places in Canada and the United States bread is baked, or, rather, dough is induced to rise, by means of a mixture known as "sourdough" a term familiar to readers of western fiction; often used of trappers and prospectors as a synonym for "old-timer". There are several slight variations, but one way of making it is to boil a large potato to the consistency of potato soup and add a compressed yeast cake while the mixture is still warm. Then stir in flour until the consistency of a thin batter is obtained. In a cool place this mixture will lie dormant and can even be frozen without killing the active principle in it. Warmth induces it to "work".

When baking bread more flour and water is added to this mixture until a large bowlful is acquired. This is placed in a warm spot overnight, and in the morning a pint or so is taken out and kept for use next time one bakes. The remainder is thickened with flour after which the procedure is normal. This mixture has a sour taste which can, however, be neutralized by the addition of a little bicarbonate of soda to the batter. It makes very good bread, and, if one bakes once a week, the process can be continued indefinitely. Sourdough, so far as I know, is immortal. But, if one bakes less frequently, the mixture becomes very sour. It takes more and more bicarbonate to neutralize this, and it is better to begin all over again with a fresh yeast cake.

It is, of course, not so easy to induce either the batter to work or the bread to rise in the open air. Tom propped it up near the embers of our camp fire and shielded it from stray draughts while we spent the morning fishing. The fish were rising more freely, and we caught twelve large Rainbow and many squaw fish. On our return we found that the saucepan of batter had overturned. This may or may not have been unfortunate.

In the afternoon we bathed, but found the water too cold to be pleasant and soon left it for the warmth of the sun-baked sand bar. Sitting there, we washed our underwear in the river, and while we were so employed, a heavily loaded boat shot the riffle and ran aground on the bar. The owner, a Swede who trapped on the Finlay River, seemed in a hurry, and as his boat was immovable, we fetched a shovel and dug it out. He offered us a tow down the long reach ahead, but we were not ready to break camp and there was still the evening rise to come, so we thanked him and resumed our washing operations.

Of all the chores I dislike most washing clothes, particularly the thick woollen underwear we wore winter and summer alike. Indeed, my washing became largely a matter of faith. I followed scrupulously the instructions on the soap packet, and my underwear was then "deemed to be clean". True, it remained a dirty grey in colour, not even as white as white, but the makers of the soap at any rate guaranteed it to be clean. Tom was of a different opinion and had a rival technique which in action resembled the treading of grapes.

The evening's fishing was not exciting, and the next morning we broke camp. Feeling, perhaps, that we were taking life too easily, we paddled continuously for ten hours. Down the long reach we had a head wind and the going was hard, but this was followed by some miles of swift water culminating in the Long Riffle, a pleasantly exciting run. Then the spruce gave way to small willow, and the river seemed to lose interest in itself. Through mile after mile of muskeg it wound round and round like a snake, and Coffee Pot Mountain, shaped curiously like a sugar-loaf with the tip cut off, was now in front, now behind us. There were many back-channels, only distinguishable from the main stream by the absence of movement in the bubbles on the surface of the water.

Then it started to rain. It was seven o'clock before we emerged into Davie Lake, and by that time we were thoroughly sick of the Crooked River. This lake is four miles long, and we paddled across it. We failed to find the outlet at the first attempt, and towards nine o'clock landed on an island and made camp, rather wet and having had nothing to eat for fourteen hours. We were about forty miles from Summit Lake.

We breakfasted leisurely off trout, rice and apricots, and did not break camp until early afternoon. Then we proceeded to explore, for we had no idea where to find the outlet. We paddled round bay after bay, and soon it became clear that we were going in the wrong direction, but the sunshine and the absence of wind were so pleasant that we decided on a circumnavigation. Once we sighted three moose, bull, cow and calf, but failed to get nearer than a quarter of a mile to them. Later, on an island, we saw some kind of stuffed waterfowl on a post guarding the door of an Indian's trapping cabin. We saw no sign of trout, and it was evening when, having explored nine-tenths of the lake, we found ourselves opposite our old camp. We paddled across and spent a second night there.

We knew now where to find the outlet, and the following morning headed towards the only bay which we had not explored. A thunderstorm accompanied by a strong head wind blew up as we were crossing the lake, and we had great difficulty in rounding the last headland before we entered the river. In the end we made it and, paddling until early evening down a more interesting stretch of water, reached Red Rock Lake where we camped on an island.

After dinner that night we were visited by a trapper who had a cabin nearby. He was a Russian by birth, an interesting man who was trying to breed marten in captivity. Like everyone else at that time, he was unsuccessful. This, I was told some years later, was because the male was segregated from the female immediately after the birth of cubs for fear he might destroy them, and was kept away too long. The previous winter this trapper had caught a timber wolf alive and unharmed, and he had it in a fur-pen when we saw it the next day. It was his ambition to cross this wolf with a husky. In this, too, he was unsuccessful but he drove it for two winters in a dog team with huskies until, finally, it escaped. Shortly after he left us, a violent thunderstorm blew up, but we were well-sheltered and very comfortable in our tent.

The rain ceased during the night, and the wind settled to a stiff breeze. After visiting the trapper and inspecting his wolf, we paddled in the teeth of the wind across Red Rock Lake, Kerry Lake, and down long stretches of almost dead water. Our wrists were beginning to swell with strain and mosquito bites, and six hours paddling was enough. Later, we learned that before the invention of the outboard motor, heavy freight boats going north were usually rowed across these lakes at night when the wind, usually adverse, had fallen.

There is a certain definite quality about the Canadian weather. As a rule, it is either wet or fine, hot or cold; there are no half measures about it. By contrast the next day was fine and calm, and the river interesting with a series of riffles to shoot. We reached McLeod Lake early in the afternoon, and camped on the first island we saw. This lake is fourteen miles long; too long, we thought, to tackle so near nightfall in our small boat; but the next day we paddled down it easily enough, having by some freak a following wind. At the northern end of the lake is Fort McLeod, the first trading post to be established west of the Rockies, and a settlement of the Pack River Indians.

The method of trading that The Hudson's Bay Company originally adopted after its foundation in 1670 was both simple and inexpensive. It built forts on the shores of Hudson's Bay—real forts, which were from time to time captured by the French—filled them with trade goods, and waited for the Indians to come to them. The independent traders from Montreal competed by travelling by river and lake to the Indian settlements—a method that was more expensive but had the advantage of contacting the Indian before he made his annual trip to the Bay.

In the end the Hudson's Bay Company had to use the same method to compete, but in the early days most of the exploratory trips were made by the North West Company of Montreal and its predecessors.

All the country to the west of the Rockies, from Alaska to the Columbia River, which now forms part of the boundary between the states of Washington and Oregon, was known to the early fur traders as New Caledonia and was opened up by the North

West Company, years before John Jacob Astor's[15] men appeared on the scene. When the Canadian boundary was drawn, we pointed this out to the Americans. They replied that we did not settle it, and that they did. This was true. The Methodists in the east, worried about the souls of the Indians in the west, decided to convert them. Their missionaries had difficulty in finding the Indians who were nomadic, but they did find some excellent agricultural land and turned themselves into farmers. The American argument carried the day, assisted it is said, by the attitude of our chief negotiator. Finding that the salmon in the Columbia would not rise to the fly, he concluded that the country was not worth having anyway.

The traders were only interested in fur. It was in 1793 that Alexander Mackenzie made his historic trip up the Peace River, on to the head waters of the Parsnip, over the height of land to the Fraser, and then west to the Pacific coast. It was not until 1805 that Simon Fraser and John Stuart left the Parsnip and travelled up the Pack River to the lake, which they called McLeod after Archibald Norman McLeod, one of the partners in the North West Company. In 1808 Fraser and Stuart made a second trip, and this time they navigated most of the river which they named after Fraser. Stuart Lake was named after his companion.

Shortly before the amalgamation of the North West and the Hudson's Bay companies in 1821 it was estimated that New Caledonia was the most valuable part of the territories, which the North West Company exploited. Sir George Simpson[16] in this report to the Governor of the Hudson's Bay Company at that time remarks:

"The present standard of barter is very high, in the ratio of 20 beaver skins for a short gun, 10 to 15 for a copper kettle according to size, and 1 for 3 inches of tobacco so that the trade must yield enormous profits; credits are as yet unknown among them and should never be introduced as they tend to make the Indians dishonest and indolent."

An obligation to explore the country was embodied in the Hudson's Bay Company's charter, but, in fact, these tremendous journeys were not only carried out by private enterprise, but inspired solely by the profit motive coupled with the spirit of enterprise in the men who made them. The two are married as naturally as government service and enterprise are divorced. The rewards—even to shareholders—were not great. The hardships to those in the field were immense. Peter Skene Ogden, a Hudson's Bay Company's clerk, writing in 1837 about a particularly tough expedition, remarked that a few years of such hardship would make a young man sixty years old, and compared the life of his trappers unfavourably with that of a convict at Botany Bay. But, he added, "They are happy."

One might expect that all this activity would have left its mark on the country, but the forests are still virgin. The canoe leaves no trace of its passing on the waters, and the teepees of the Indians leave little on land. We had been travelling through a country with a history, but with no ancient monuments.

Here at Fort McLeod we came upon the first of them, a trading post, although not, of course the original building, which had been in operation since 1805, the first to be established west of the Rockies, or entitled, at any rate, to share that distinction with Fort St. James on Stuart Lake.

We made camp that evening on an island about a mile from the trading post and were soon visited by a boat full of Indians from the settlement. They had some curiosity about us, but had really come to find out if we had seen their priest. This settlement had a church but no resident priest. When the winter's trapping had been good, and the Indians felt that they could afford it, they sent out a party in the spring to Prince George to fetch a priest, and this year he was somewhat overdue.

He arrived the following day. Seated in the middle of a large boat propelled by an outboard motor, dressed in black with a black umbrella over his head, the priest looked to us faintly sinister, as if he were coming to celebrate funerals rather than weddings. Not so, apparently to the Indians, for a crowd of boats put off to escort him, and when he landed, there was much firing of guns. These Indians have a bad reputation as thieves, but they certainly did not appear to be irreligious.

Some years later I talked with an old Indian chief[17] at Fort McLeod. Two members of the Royal Canadian Mounted Police had recently made a journey of some hundreds of miles in order to take three Indian children out of the bush to be educated. The chief said; "Priest good, police good. Before priest and police come, much fighting among Indians. Now no fighting. School bad. Teach Indian to trap; school no good for him." For different reasons he took the same view as our taxi driver in Prince George, and it is worth remarking that Sir George Simpson, too, was privately of the opinion that "An enlightened Indian is good for nothing."

I do not know what Gods the unconverted Siwash worship, but they have a custom that has some affinity to Christian teaching. At times an Indian will give away all his worldly possessions and embrace Holy Poverty. The church cannot well condemn this practice, but the Indian Agency, upon whim the devotee becomes a charge, can and does. One policeman, ordered to put a stop to this ceremony, found an Indian had tied all his possessions to the branches of a spruce tree around which he was solemnly dancing. "Don't you know," said the policeman, taking him aside, "this potlatch is against the law?" "This no potlatch," said the Indian, "This all same white man's Christmas." The policeman had a sense of humour and no orders to stop a Christmas party. He rode away, and the ceremony continued.

When the excitement of the priest's arrival had subsided, an old trapper approached our camp in a beautiful dugout canoe, driven by a three hp outboard motor. It looked a tricky craft to handle. This trapper was known as "seven-fingered Gus", for he had lost three fingers of his right hand in a trap and retained only the thumb and trigger finger. He had come into the country in 1907, and it was several years since he had been nearer civilization than Fort McLeod, for civilization to Gus meant rum and rum meant trouble. He found it easier to cope with life in the bush, although that, too, had its troubles, and he complained that on his trap-line the grizzly bears kept coming "too damn close," much as a city man might complain that his train was always overcrowded. The previous winter one had come closer than usual. Turning from fixing a trap, Gus had found the bear behind him, just rising to strike. "And then?" we asked. "Shot him through the eye," said Gus. "Knocked him cold." A nervy and a lucky bit of shooting. It is not often that a bear, black or grizzly, will drop in its tracks.

There is a great deal of work in skinning, cleaning, stretching, and packing a bear hide out of the bush, and practically no sale for it, so, as a rule, trappers shoot grizzly only when they have to—which is seldom—and treat them as vermin. But this pelt was exceptionally fine, and Gus had brought it out. He wanted seventy-five dollars for it, but to his great disgust, Fort McLeod only bid him fifty.

In the afternoon we broke camp and paddled over to the Hudson's Bay post on the mainland. Here we bought a few luxuries: a can of milk, one of strawberry jam, one of pineapple, and some more tobacco. Then we inspected the bearskin. It was large, very fine and well furred, a yellowish brown in colour. We sympathized with Gus; when one considered the price of a rug made from it, the raw pelt should certainly have been worth more than fifty dollars, for prices were then high. Later, we learned that he took it out to Prince George hoping for a higher bid, but the attraction of the liquor store again proved too much for him and, in the end, he sold it for less than the Hudson's Bay Company had offered.

The Crooked River is only ten to twenty yards wide, and in one place is sometimes dammed across by beaver during the fall. The Pack River is twice as wide, but, apart from its greater width, a very similar stream. We started down the Pack late in the afternoon and after shooting several small rapids, ran into a thunderstorm and drew the boat under some dense willows to shelter. These turned away most of the rain until a cloudburst occurred which quickly soaked us to the skin and kept us hard at work bailing the boat. I have never seen so much water fall in so short a time. When the rain had settled to a steady drizzle, we slide out into the stream and soon reached Trout Lake. This lake is three miles long, but proved rather longer for us as our first attempt to find the outlet failed. We saw no sign of trout, and were not tempted to wet a line.

Almost immediately on leaving the lake one comes to Cross Rapid where the water rushes in two streams from either bank to meet in the centre of the river. We shot this successfully although we missed the best channel—the only one in low water—which lies very close to the north-west bank. It was now raining heavily again, and we swung the boat into the eddy at the foot of the rapid and looked for a place to make camp. The undergrowth was very wet, and the most important thing for us was a large fire. We found a fallen spruce, and built one against it; there was no danger of it spreading that night. When the fire was well ablaze, we put up the tent, stripped, and standing naked in the rain, dried out our clothing, threw it piece by piece into the tent, crawled in after it and dressed again. Feeling disinclined to leave shelter, for we had no outer clothing of any kind except shirt and trousers, we supped of strawberry jam, canned milk, and ships' biscuits.

The rain ceased before dawn, and the next day was very fine and hot. We declared it a holiday and went fishing. Here we caught our first Arctic trout. The Arctic trout is really a grayling with the characteristic tough, scaly skin, dorsal fin, and bone structure. The head, too, is similar, although it always seemed to me that the mouth was rather larger than that of the English variety. But this impression may have been due to the fact that the fish themselves averaged rather larger than those I have caught in England. This fish rises to the fly, is a game fighter, and excellent eating. We had great sport that day, but it was too hot for comfort. As the heat increased, many garter snakes came out of the rocks to sun themselves. This is the only snake to be found in this part of Canada and is quite harmless.

The mosquitoes were not so harmless, and the river was not wide enough for us the escape from them by fishing from the boat. The following day was hotter still, and we were glad to break camp. The Pack River is only nineteen miles long, and after shooting several small rapids, we reached the Parsnip River early in the afternoon; a large river at last, with a strong current in many places. Before entering the Parsnip, we let the boat swing idly in an eddy under a high cut-bank, and ate our can of pineapple. It tasted very good.

Our days of hard paddling were over. There were no more lakes to cross with the inevitable head winds, and, as there is only one rapid on the Parsnip, scarcely noticeable except in low water, over a hundred miles of easy navigation lay before us. It was hardly too dark to travel on the river at ten o'clock at night, and light again a half past two in the morning, so we started early and ran late to avoid the flies.

The waters of the Parsnip are green in colour and in high water full of sediment which whispered against the sides of the boat as we drifted along. From time to time we landed on sand bars and panned them. Usually we found colours, but nothing that looked interesting. The bars nearer the source are said to be richer, and Gus had told us that they had not been washed for five years, but we had no intention of going upstream under our own power. We allowed the boat to drift as the current took it, broadside on or stern first, so long as we could see good water ahead. We met nothing on the river except a beaver who must have been looking for a new home. He took no notice of us until we paddled towards him, then, with a warning splash of his tail, he dived. All the way down we saw from time to time the colourful flash of kingfishers near the banks, and in the air above hawks hovering and swooping, and occasionally a bald-headed eagle. We had no luck at all fishing; the water was too heavily coloured. So we lived on rice, beans and bacon; an odd diet in a temperature of ninety-five in the shade.

The Parsnip is named from the wild parsnip growing on its banks, the Crooked from the nature of its course, the Finlay after its discoverer, and the Peace because two Indian tribes[18] had made a treaty there. Most of the names that have not been anglicized have the suffix "ka", meaning "water", as, for example, Ingenika, Ospica Omenica—all tributaries of the Finlay. One place name, however, deserves special mention, Deserters' Canyon on the Finlay River.

Named originally to commemorate the desertion of the canoe crew of an early explorer Samuel Black, the following incident gave it an added significance. Two men of Danish extraction were trapping on the Finlay in 1916. Feeling no interest in the war then raging in Europe, and fearing conscription, they arranged with a fur buyer in Prince George (a Greek, who shared their views), to come to Finlay Forks every year in the spring with a boat load of provisions for them, so that they need not come out of the bush. This the fur buyer did. He met them at Finlay Forks in 1917 and 1918, and, in answer to their enquiries, told them that the war news was very bad, which was true, and that there was hardly any price for fur, which had been true enough at the beginning of the war, but was true no longer. However, out of the goodness of his heart, he agreed to take their catch in return for the provisions he had brought them, and they were in no position to bargain with him.

In the spring of 1919 he told them that the war was going from bad to worse, and that, as every available man was being conscripted, it was hardly wise for them to come as near civilization as Finlay Forks. He suggested meeting them at the foot of the canyon known as Deserters' Canyon, and this he did in 1920 and 1921. In the spring of 1922 he did not arrive, and the two trappers, having no food, were obliged to make their way to the Hudson's Bay post at Whitewater. They entered in some trepidation and made their usual enquiry about the war. "Good God!" exclaimed the store keeper, "Has somebody started another war?" When the truth came out they were laughed out of the country, but they did not catch up with the fur buyer. He had preceded them in a state of considerable affluence.

During the afternoon of our third day on the Parsnip the river increased to about half a mile in width, and, as we drifted along close to the east bank, Tom saw a cabin and what appeared to be a tin sign overgrown with creeper. He suggested landing, but I was too sleepy to listen until he remarked that the current was very swift, and, if we did not make a landing soon, we should be round the bend of the river and into whatever lay beyond. I roused myself, and we drew into a small eddy, and went back to look at the sign. It read "Dangerous Rapid". A man came out of the cabin. He was packing in supplies for the Hudson's Bay Company, and we learned from him that we had nearly drifted into the Finlay Rapid, half asleep, and that asleep or awake, we could not possibly run the rapid in our boat.

There was a small store on the opposite bank—the north bank of the Finlay—but, as the current was very strong, we could only reach it by poling half a mile back up the Parsnip and making a long detour. While we were debating what to do, the store keeper crossed over and, lashing our boat to his, gave us a tow across. Then he gave us a large meal of potatoes and bully beef. The potatoes tasted very good after a fortnight without fresh vegetables of any kind, and I said so to our host. "Nothing's too good for me," he declared, "if I can get it."

Louis Peterson, for that was his name, was store keeper, fire warden and, as he remarked, "every Goddamn thing"; seventy-four years old and half blind with cataract, but quite unsubdued either by age or infirmity. He had done many things in his time, proving-up a homestead in the United States and prospecting in the Yukon in ninety-nine among them. He did not make his fortune in the Yukon. Seventy dollars a-day, he told us was the most he made, and that was not enough. The cost of living ate it all up. He had also prospected up the Finlay River and its tributaries, but now he was content to trap a little around Finlay Forks in the winter, and run his small store and grow a few vegetables during the summer months.

Most men of his age have one fad or another to which they attribute their length of life, but Peterson said cheerfully that he had never taken much care of himself, adding, as confirmation, that he had "been in the saloon business twice." But he had some good advice to give us; like most good advice, the fruit of his own mistakes. He had served nine years in the Danish army, and he warned us never to take up soldiering. He had a wife living somewhere, he believed, and he cautioned us against matrimony. And he was strongly anti-clerical, although the reason for this did not appear. It was a simple philosophy. You found some place where there was no government to conscript you, no woman to marry you, and no priest to damn you, and lived happily ever after.

He had three ambitions: to live to be a hundred years old, and sell his place at Finlay Forks remuneratively (he had a hundred and sixty acres of land there), when a railway was built through the country and make a trip round the world.

He achieved none of them. Two years later a railway survey was made, but it missed Louis' place, and was never built. Three years later Louis died. He kept going to the last, although his sight was so bad that he had to have a hand-line to guide him to and from his water-hole. In the end he dropped in his tracks and died. Despite the strenuous nature of his life, his epitaph, if he had had one, might have been that of Regnier[19]:

> J'AY vécu sans nul pensement,
> Me laissant aller doucement
> A la bonne loy naturelle,
> Et si m'étonne fort pourquoy
> La mort osa songer à moy,
> Qui ne songeay jamais à elle.[20]

He had no religious beliefs that I know of, but it will take a lot of rough country to keep Louis out of paradise if the streets are paved with gold.

As for his view of life, I agree, now that I can speak on equal terms in the matter of age that "taking care of oneself" is no part of my philosophy. But, since all undue excitement either of body or mind is inevitably followed by a relapse which, beyond a point, I find intolerable, my way of life may well be more "Godly, righteous and sober" than was Peterson's. For the rest, the solution of the problem can hardly be to empty it of any significant content. I think I should express my philosophy in terms of the habits of the river trout—in parable, since "to them that are without all these things are done in parables," and, as Izaak Walton said of his recipe for cooking pike, it is "too good for any but anglers, or very honest men."

The river trout feeds largely on what is brought down by the stream, and in any case, has to lie facing the current to avoid drowning. This can be exhausting, so he likes to lie in the slack water of an eddy, behind a rock or near the bank, and snatch his meal as it passes in the current. If there is room for more than one, the largest lies at the head of the pool on the dividing line between the eddy and the main stream, the next largest below, and so on. The angler knows this, and casts his fly on the stream just above the pool. If he catches the largest trout, the next in size moves up. I conclude that, on the whole, the position of second vice-president is the one to aim at, and one should avoid promotion.

Peterson advised us to acquire some land at Finlay Forks, as a railway was sure to be built through the country soon. The idea of a railway giving to the grain grower in the Peace River Block a direct outlet to the Pacific coast at Prince Rupert seemed to be in everybody's mind. To trappers it meant little but trouble, but, like Peterson, they were determined to make what they could out of it, and then, if it spoilt the country for them, to move further north. But it seemed unlikely that a railway would be built for the sake of the wheat farmers. If a large body of gold ore were discovered in the mountains to the west, that would be another matter, and any such discovery would determine the route.

In the morning we examined one or two lots which happened to be surveyed. They were covered with burnt-over spruce; poor land which held little interest for us. We turned our attention to the Finlay Rapid. Peterson had left early in the morning to investigate a haze up the Finlay which he thought might be caused by a forest fire, but, before he left, we had questioned him about the rapid. He told us that, if we were determined to run it, there was a channel near the north bank which would be possible if the water were high enough, which was doubtful. Failing that, we should have to keep well out near the middle of the river where there was rough water, but little danger from rocks. The best channel was near the south bank, but it was very narrow, and a tricky piece of navigation. If we attempted it and missed it, we could hardly hope to come through.

One difficulty in handling a boat such as ours in swift water and without an engine is that one has to make up one's mind where to run into a rapid a quarter of a mile or more before one comes to it, and at this distance one cannot get more than a general idea of the water. Manoeuvrability is small; there is no turning back, and the only safe way is to make a landing while it is still possible, walk down, and investigate. We did not do this, and, as we approached the rapid, we could only see half of it, for there is an island in the middle of the river. I stood up in the bow, and looked at the water. The inside channel on our left was full of rocks, and it was impossible to see if there was enough water to get through. I decided we must keep well out in the river; water, however rough, is preferable to rocks.

There were plenty of rocks in the middle of the river, but I judged these to be well below the surface with the exception of a few large ones which stood far out of the water, and I knew that, under our own power, we could not run onto one of these if we had tried. We picked the largest, which stood sheer out of the water and seemed to have a good channel alongside, and made for it, paddling hard to keep steerage-way. A few yards from it we stopped paddling, and the boat headed straight for the rock until, seized by the powerful throw-off, it shot neatly into the channel and down into the rough water, but, as Peterson had said, no rocks near the surface. We should have had no trouble in a suitable boat, but ours had not been built for this. The bottom was not only flat, but dead level; not an inch of rake fore or aft. Whenever it plunged into the trough of a wave it tried to hit the bottom of the river. We shipped water, and had to bail, but otherwise came through without difficulty.

We were on the Peace River now, and in the heart of the Rockies, but we could see nothing of the mountains, which were shrouded in heavy mist. Presently it began to rain, and we decided to look for a place to make camp. We saw an old cabin on the south bank and started to make for it, but the current was too strong for us. Almost immediately, as we passed the mouth of a tributary stream we saw another cabin on the north bank, and landed just below it. We were rather wet, but the cabin possessed a good stove, and we soon dried our clothing and supped off cornmeal cakes, beans, and bacon.

This was our first night under a roof since leaving Prince George, and some old magazines there were our first reading matter. They featured, I remember, a story about Baron von Richtofen, "The Red Knight of Germany"; in addition to the magazines we spent some time reading remarks and signatures scrawled on the cabin door, of which there were a great number. These ranged from "Took two carrots", signed by a neighbouring trapper, to "Put through the Peace-Pacific railroad", signed "General Sutton". Sutton was the one-armed Englishman who had been Chief of Staff to Chang-Tso-Lin, the Manchurian war lord. He had left for London, we were told later, to raise capital for this project, but London, I fancy, had not forgotten the Grand Trunk Pacific.

In common with all trappers' cabins, the door of this was never locked, and its shelter and such supplies as it contained were at the service of any traveller to the extent of supper, bed, and breakfast. A pleasant, indeed a necessary local custom, only possible in a country well away from civilization where no parasitic riffraff could exist. In return the traveller was expected to leave plenty of shavings and chips to start a fire. One always did this in winter since a man with badly frozen hands might be unable to make shavings, and, if he could not start a fire, would freeze to death as easily inside his cabin and outside it.

During the night we were visited by a porcupine, and Tom argued with it for quite a while before it consented to go away. There is a prejudice against shooting them, as a porcupine is the only animal that an unarmed man lost in the bush can kill. Trappers, however, kill them as vermin, for if a porcupine comes upon an axe, for instance, it finds the salty flavour of the haft where it has been handled so attractive that it will eat it away. Like the skunk, the porcupine's armament is purely defensive, but so efficient that he is usually left unmolested by other animals.

The porcupine has the dignity of being incorporated in an Indian legend of the Flood. One of the Gods, so the story goes, was drifting about on a partly submerged log when a porcupine swam up, and, with the customary insolence of its kind, appropriated

Mt Selwyn as seen from the Wicked River.

the dry end of the log, forcing the God to sit in the water, a fisher swimming by was so enraged by this lack of respect for divinity that he attacked the porcupine and drove it off the log. The body of the fisher was full of quills, but the God plucked them out, saying "so shall it ever be." And so, according to the legend, it still is, and the quills of the porcupine which will slowly work their way into the body of an animal, festering, and often causing death, are harmless to the fisher. This story is edifying, but, unfortunately for the fisher, it is not true to fact. During the following winter we caught a fisher, and after skinning him, found several quills embedded in his flesh. He still fights the porcupine, but no longer, it would seem under divine protection.

In the morning we were glad to leave the stuffy cabin, much as we had welcomed its shelter the night before. Most of the clouds had cleared away, but heavy banks of white mist still lined the mountain sides. The scenery was rugged, and, among many smaller peaks, about a mile to the south east Mount Selwyn rose sheer from the river some four thousand feet. Selwyn is sometimes called "The Mountain of Gold", and is said to contain plenty of ore but of too low a grade to pay for working without better transport facilities. A hundred yards above the cabin a river joined the Peace, and this, although we did not know it, was the Wicked River about which we had heard in Prince George. It is ten miles from Finlay Forks, but the map, which was very inaccurate, showed it as more. We thought this stream must be Anderson Creek, and did not explore it at this time.

We set out to find a strong head wind, stronger, in fact, than the current of the river at this point, and we had to paddle for a while although we were going down stream. It was a delightful morning. The snow on Mount Selwyn came nearly down to river level, and the breeze from it kept us cool. The mountain peaks towered over the banks of mist, and we seemed to be floating on the top of the world where no river ought to be. We passed close under the mountain and saw a mining outfit encamped at its foot. After running through some rapid water we came to the mouth of a tributary stream on our left and decided that this must be the Wicked River, although we had now come much more than twelve miles from Finlay Forks, and here we made camp. It was, in fact, Barnard Creek.

The next morning was wet, and we lay in the tent, smoking, but about midday the rain cleared, and we set off up the stream to look for the fishing pool. The trail, good enough, no doubt, in winter with three or four feet of snow on the ground, was very difficult—a tangle of windfalls with little sign of cutting—and the mosquitoes were fiendish. About four miles up-stream we found a camp site and a small pool in the river, but it was clear that this was not the pool we were seeking. Our faces were covered with blood and flies, and we bathed them in the stream and made a smudge-fire. After resting for a while in comparative comfort we fished for a few minutes and caught three—one Arctic, one Rainbow and one Dolly Vardon trout. The only real trout was the Rainbow; the Arctic is a grayling, and Dolly Vardon a char. Like the Arctic, the Dolly Vardon is very good eating. It grows up to ten or twelve pounds in weight and, salted and smoked, it makes a very good substitute for smoked salmon. Its only vice from the fisherman's point of view is that it seldom rises to the fly.

We continued half a mile further up the creek, then, deciding finally, that this could not be the Wicked River, turned back. Trying to take a short-cut, we found ourselves in the middle of an ancient swamp full of the corpses of fallen trees, much larger than any growing thereabouts today. They looked sound enough, but, when we tried to walk along the trunks, we trod right through them, and fell among "Devil's Truncheon," a weed which grows in swampy places. It is covered with sharp spikes, and its scratch is mildly poisonous. Mosquitoes settled on us in swarms, and when we reached camp, we had scarcely sufficient self-control to light a smudge-fire. It was a particularly bad year for mosquitoes on the Peace River.

The following day was Sunday, and we made it a day of rest. On Monday we were delayed by a thunderstorm and made a late start. We landed about a mile downstream to look at an old cabin, and here Tom shot a grouse neatly through the head with the .22. Further down we landed on a bar and washed some gravel. We found plenty of black sand, but no gold, and here we forgot our pick when we left. Shortly afterwards we came to a river running into the Peace on our right, and another about a mile below it. We knew these must be Point Creek and the Clearwater River. We were about thirty-five miles from Finlay Forks, almost in the middle of the range of the Rockies, with spruce-covered mountains coming steeply down to the river bank. When we discovered an apparently inexhaustible supply of large Arctic trout at the mouth of Point Creek, we decided to camp here for a while, and later, perhaps, pole back to the Wicked River. We cooked the grouse in the hot sand underneath our camp fire that night, and had a very pleasant supper.

We spent seventeen very happy days at Point Creek, sixteen of them fine, which means a great deal under canvas. A hundred yards or so downstream was the headquarters cabin of Jack Pennington who trapped the Clearwater, and it contained a few welcome copies of "Punch"[21] for the year 1925—so old that they were new to us. Some distance back in the bush was a derelict cabin which had been built by two prospectors who travelled by this route from the Yukon to Edmonton after the gold rush of ninety-eight.

Wicked River falls.

I had once met a very old prospector who had known these men. He told me that they had made a strike either on the Clearwater or on Point Creek. One of them, he said, had died there of scurvy during the winter; the other, after reaching civilization, had died also before he could return. No rumour of this strike survives locally, and this was one of the few places where we found no sign of gold—nothing but black sand. Perhaps we did not search hard enough, for neither of us, I fear, is of the stuff of which true prospectors are made. This "stuff" consists chiefly of a boundless credulity, and a willingness to do any amount of hard work to verify any old wife's tale.

It is said there are no prospectors in Heaven. They are too credulous even for paradise. The story goes that the first prospectors who applied for admission with the necessary credentials were allowed to enter. When they saw that the streets were paved with gold, they decided that there must be a very big body of ore somewhere, and started to look for it. Their diggings made a nasty mess of the scenery, the authorities disapproved, and the next prospector who arrived was refused admittance. After some consideration he suggested to St. Peter that he would undertake to persuade all the others to leave if he might remain, and St. Peter agreed to the bargain. So he entered paradise, and finding a miner busy digging a hole, took him aside, and told him in strict confidence, that a big strike washing out at seven and half dollars a pan, had just been made in Hell. The miner thanked him, packed up hastily and left in the direction indicated, as also did his companions when they were let into the secret. When none was left, the originator of the rumour stripped off his pack and, picking up his shovel, prepared to start work when his eye fell on the last miner hastening out of the gates of Heaven. "By God", he said, "there may be something to it after all" and he, too, made his way out of paradise.

By this time we both had a craving for fresh meat. Louis Peterson's bully beef and the grouse that Tom had shot had been the only additions to our diet of bacon and salt pork for nearly a month, apart from the trout we had caught. According to the strict letter of the law we were not allowed to shoot anything at this season unless we were actually starving, but we felt that, in our situation, shooting for the pot is always allowable, so long as the animal is not protected on account of its rarity and is not a female with young. To obtain meat at the first opportunity was now our intention.

One morning, while washing the breakfast dishes in the creek, I saw a black bear enter the river from the opposite bank, and start to swim across. He had evidently allowed for a drift of several hundred yards, and intended to land opposite Jack Pennington's cabin. I hurried back to camp with the news, and Tom picked up the Winchester and took cover behind a boulder on the shore. I had the choice of the .22 or the axe. The .22 would have been quite ineffective, but a three-pound axe is a useful weapon, so long as a bear remains on all-fours. I chose the axe, and took up my station behind a tree where I could intercept the bear if Tom only succeeded in wounding him. My thoughts slipped back to an early morning years before in eastern Poland when I had waited with an unaccustomed weapon for a wild boar.

Now, however, I was only a reserve, and did not expect to be called upon. Nor was I. Tom opened fire when the bear was some twenty yards from shore, and the bullet passed close to its heart. The bear reared, shook himself and came on, the water crimsoning about him. The vitality of a bear is tremendous; he will often travel twenty or thirty yards when shot through the heart. This one came out of the water, staggered up the beach, and would have reached the bush had not a second shot broken his spine.

He lay with his paws crossed in front as a dog will often lie, and we both felt rather sorry for him. But we also felt hungry, and lost no time in skinning and butchering him, taking out his liver first and putting it to cool in the creek. Meat we now had, but, although we went over every inch of him, we could not find enough fat to fry a steak. The bear had been hungrier than we were. He was quite healthy as, indeed, bears usually are. Deer and moose, when their natural enemies the wolves are reduced in numbers, frequently become tubercular no doubt because the weaklings are no longer killed off, but bear both black and grizzly, always seem to remain healthy, although the black bear sometimes become quite thick on the ground.

We cooked our bear steaks without fat, heating the frying pan red hot, and then sprinkling salt on it before putting the meat in; not a bad way. Then, encouraged by our well-fed feeling, we decided to pole back upstream and recover our pick from the bar where we had left it. It took us an hour and fifty minutes to pole up, and ten minutes to return. We abandoned any idea of poling back to the Wicked River and spent the next few days drying and smoking the bear meat, and building a small cache. The mice had become so bold that Tom could no longer discourage them by sitting up at night and shooting at them with the .22 by firelight. We built our cache with small poles and roofed it with birch bark. It served our purpose, and, indeed, remained standing for some years.

We also explored a few miles of Point Creek and the Clearwater River. We found a pretty waterfall some distance up Point creek, and the pool below it contained large, pink-fleshed Rainbow trout. A few miles up the Clearwater was a canyon containing several pools, one of them very deep. Here were Arctic, Rainbow, and few very large Dolly Vardon trout. The rocky banks were so steep, and the water so clear, that one could look down and see the fish lying beneath each other tier upon tier, almost, it seemed, to infinity, and watch every detail of a rise. I noticed that, after catching a few in one place, others would rise and look at the fly without taking it. A change of fly would deceive them again. The nature of the fly hardly seemed to matter, although I never tried running through a large selection.

We had been in camp at Point Creek for twelve days when our peace was shattered by the sound of an outboard motor. It proved to be Jack Pennington coming upstream to provision his cabin for the winter's trapping, and with him was Jim Beattie, who trapped further down the river, and who, we had been told, had a ranch there. It was pictured by trappers thereabouts as a land flowing with milk and honey, and Jim himself was a legend even in his lifetime.

We learned that in five days' time Louis Peterson was to meet Jim at the head of the Ne Parle Pas rapid with a ballot box, for it appeared that an election was taking place, and Peterson was collecting the few votes—about a dozen, I think—from the Finlay River. Jim was to take the box on to Hudson's Hope. He suggested that we should come down with Peterson, and offered to give us a hand down the rapid and a tow downstream, if we wished. This seemed a good plan, and we accepted it.

Jack Pennington gave us some potatoes, a welcome addition to our larder, and the next day I ate the biggest meal I have ever eaten. In the morning we went up Point Creek to the pool which contained the pink-fleshed Rainbow, and caught fifteen of them. None would weigh less than half a pound and the four largest about a pound and a half each. In the evening we cooked these four under the campfire, gypsy fashion, fried the rest, and ate the lot. We also ate at this meal eighteen potatoes, three rashers of bacon, and some strawberry jam.

Afterwards, Tom, whose stomach sometimes failed him in moments of heroic achievement, rolled about in some discomfort calling in vain for cherry brandy, his usual specific on these occasions. I should have welcomed half a pint of cherry brandy myself, but my sense of satisfaction outweighed my physical discomfort. I am fond of trout, and had often thought that one day I would enjoy some such orgy. But, the following morning for the first time in my life, I ate trout for breakfast without relish.

Afterwards we prospected for three days, although Jim had confirmed our suspicion that there was no gold hereabouts. We discovered none, and the legend of the strike made by the two prospectors from the Yukon remains a legend.

Louis Peterson arrived with the ballot box earlier in the morning than we had expected, and we had to break camp in a hurry. We loaded up hastily, lashed our boat to his and said good-bye to Point Creek, but our departure was an anticlimax. His engine refused to start. Louis "goddamned it" freely, remarking that he had twice nearly drifted into the Finlay Rapid while trying to start it. Eventually, after he had drained some water from the carburetor, it responded to his efforts, and we passed the mouth of the Clearwater River. We had about five miles to go to the Ne Parle Pas Rapid, and Peterson confided to us that he did not like the river. "I know every stone in Parsnips," he said, "but I don't know the Peace so well, I don't like it." We didn't wonder, for he was half-blind, and kept asking us if we could see any rocks.

We reached the Ne Parle Pas Rapid about noon. It is a peculiarly silent rapid, as its name implies, and as it lies round a bend in the river, one comes on it without warning. Jim Beattie was awaiting us. He had made the journey upstream the previous day, and spent the night there. We lunched together and then lined our boat down the rapid.

This rapid is caused by a large reef which runs at a slight angle across almost the whole width of the river from the south bank to the north. The reef breaks up into large boulders some yards from the north bank, and immediately below is a strong eddy which throws-in violently towards the shore. There is a drop of eight feet in the river bed. When running the rapid one has to dodge in among the boulders close to the north bank, and then, when one has passed the reef, turn out almost at right angles underneath it to avoid the eddy. In most water conditions this is not difficult if one has a suitable boat and engine, but power is almost essential except in very high water. We should not have had much chance in our boat, and were glad of assistance even in lining it down, for the boat had to be dragged through the eddy, and at the same time prevented from dashing itself to pieces on the boulders which line the north bank.

At the foot of the rapid we said good-bye to Peterson with the fervent hope that his engine would start before he drifted into the rapid, lashed our boat to Jim's, and set off for his place at Twenty Mile Creek[22]. After travelling a mile or so we passed the mouth of the Ottertail River on our left and avoided the Little Ne Parle Pas Rapid by taking a back channel which, although very shallow, was still navigable. This was the last of the rapids, and as Jim ran the engine, we had leisure to observe the country through which we were passing.

Below the Ottertail the mountains no longer crowd in on the river, pockets of flat land begin to appear a few hundred acres in extent, flanked by steep benches, and as we travelled these flats became more numerous and larger. Although spruce was still plentiful, poplar was to be seen more frequently. The soil was richer, and is, in fact, largely composed of silt. According to geologists, before the Peace River cut its way through the mountains the country through which we were passing had been a lake. At a more remote period it must have been under the sea, for a few years later four fossilized skeletons of ichthyosuari were found near the Ne Parle Pas Rapid.

We passed the mouth of Schooler Creek on our left, and then came to Carbon River[23], which gets its name from the coal deposits near its upper reaches, and runs into the Peace from the south. Its estuary is split into two parts by a small, spruce-covered island, and on this island lived Charlie Jones and his wife. Apart from Finlay Forks, this was the first place where we had found anyone in permanent residence since leaving

Red Rock Lake. And a woman! Obviously the first outpost of civilization. There were, in fact, curtains on the windows. Charlie and his partner, Johnny Darling came down with us to Twenty Mile[24] to pick up a boat and engine they had bought from Jim.

Looking west up the Peace River from Schooler Creek.

On the northern bank of the river four miles further down we passed Branham's Flat[25], the scene of an unsuccessful mining venture. There is gold here, not only in the river bars, but in the bank itself, and the deposit had proved rich enough to result in the flotation of a mining company some years previously, but too poor to pay for working. Four miles below Branham's Flat Twenty Mile Creek enters the Peace River on its north bank. Here we landed late in the evening and slept under a roof again in Jim's bunkhouse.

Twenty miles downstream the Peace River plunges into the Rocky Mountain Canyon[26], twenty-two miles of white water, unnavigable, and, indeed, at that time, unexplored. It emerges in the flat prairie country known as the Peace River Block. We had seen almost all the country that we had set out to see. It remained to decide whether we would settle here, or whether we should go down river to Fort St John, or even as far as the town of Peace River, go out overland to Edmonton, and strike further north the following year. Jim offered us hospitality until we reached a decision, but I think our minds were already made up. If we could acquire a trap-line and a winter's grubstake we were content to remain on the Upper Peace River.

On the Peace River in winter, looking west from Schooler Creek.

Peace River

*"Now I lyue at libertie after myne owne mynde and pleasure,
whyche I thynke verye fewe of these great states,
and pieres of the realmes can saye."*

—Sir Thomas More.

WE AWOKE IN the morning to the crowing of cocks, porridge with fresh milk, followed by bacon and eggs for breakfast, and all the atmosphere of a settled life. Looking west from the farm buildings towards the Rockies, we gazed over some eighty acres of cultivated land, most of it under wheat, oats and timothy, to a distant horizon reminiscent of the Cumberland hills. The Peace River we could not see; it was hidden by the height of its own banks; and we missed the murmur of running water, the most soothing and enticing sound in the world, which had enchanted us for so long. For a time, our travels were over. We were content to sit quietly and look around.

We were still in the foothills of the Rockies. The farm had been made on one of those pockets of land which appeared as the mountains receded from the river, and to the north, south and west there was nothing but trapping country for hundreds of miles. Transport here was by river and over trails few of which were practicable for pack horses. To the east, thirty-four miles away, lay the village of Hudson's Hope situated at the foot of the Rocky Mountain Canyon, on the extreme western edge of the Peace River Block. A wagon road, so-called, connected Twenty Mile with the village.

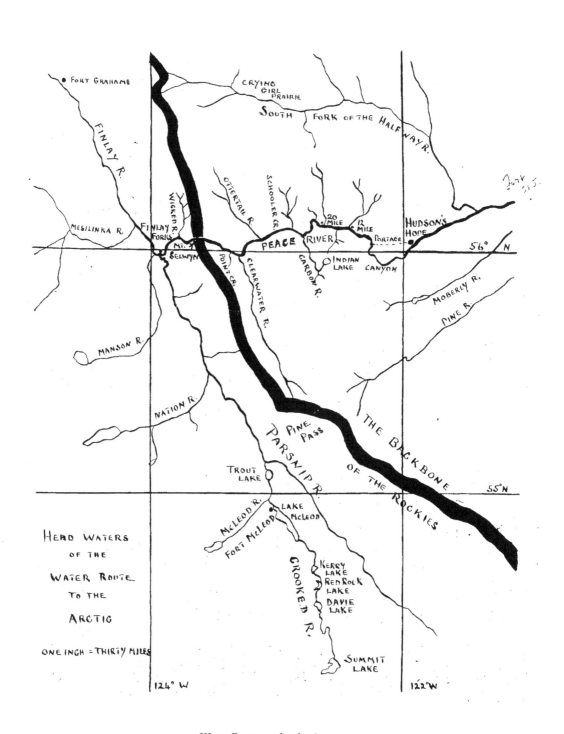

Water Route to the Arctic.

At the old boat landing of Twenty Mile, 1931 – from left: Jim Beattie, unknown, Tom Stott, Girlie (Mary) Beattie, Elizabeth Beattie, Bobby Beattie, Reg Lucas, Olive Beattie, Toulie (Louise) Beattie, Gertie Brown (Lucas).

This road ran parallel with the river—still navigable—for the first twenty miles, and had been made by Jim Beattie and other local trappers. It was surfaced nowhere, and graded only where grading could not be avoided. Two bridges had been built, but wherever possible one drove across the beds of streams—when the water was not too high. Tree-felling and clearing windfalls, still necessary after every high wind, and the marks of wagon wheels, distinguished the road from the surrounding bush.

It was subject to curious hazards. A few years later a family of beaver moved into Twenty Mile creek, and signalled their presence by felling a succession of large poplar trees across the road. Next spring their damming operations flooded half a mile of it, and, as Jim did not wish to trap them out, two miles of new road had to be cut over the hillside to by-pass the flooded section. The change which the engineering work of a few beaver can bring about over comparatively large areas can only be realized by those who have seen a country where the beaver have been trapped out and their dams cut.

The succeeding fourteen miles to Hudson's Hope was the portage round the Rocky Mountain Canyon. Jim ran the portage, personally or by deputy, at this time, and kept two teams of horses and two wagons for the purpose. As a trail it had been in existence for a long time, and now occasionally received a small maintenance grant from the state.

This road and its extension into the prairie country to the east was the chief political issue in the district. In conformity with the basic principle of democratic government the local residents voted at election time without fear or favour for the party which promised to do most for the road. At each end it rose sharply for about four hundred feet, and some of the gradients were fantastic, but it afforded a practicable connection by wagon or sleigh between the upper and lower waters of the Peace River.

In this isolated spot the sale of farm produce on any large scale was obviously impossible. Jim farmed because farming was in his blood. He had grown up in the hill country near Cockermouth[27], emigrated to Canada, and after spending some years trapping and packing during the construction of the Grand Trunk, had married and settled in the Peace River country in 1913. Fortunately, he had married a girl from North Dakota who was accustomed to frontier life, for their social life depended on passing strangers, and they were hundreds of miles from the nearest doctor. They had five children, four girls and one boy (later in 1936 twin boys), all under fifteen years of age.

Jim was about forty-five years of age, a burly, fresh-complexioned man of medium height. Although he was putting on weight, he was full of restless energy, and a craving to try out new ideas. He neither drank nor smoked, and five or six hours' sleep was the most he could tolerate, or permit anyone else to enjoy when he was around. When it came to work he was a slave-driver, both to himself, his family, and anyone else he could rope in. Between whiles he was good company, humorous, fond of good eating, and the soul of hospitality.

Charlie Jones, who had joined us at Carbon River, was a very different character. A long, lean man who chain-smoked cigarettes, he came, I believe, from Indiana and was about sixty years old at this time. He had had an eventful life, and was also deeply read in the "penny-dreadful" type of literature. A lively imagination, and a sanguine temperament acting on this groundwork made him an excellent companion for anyone with a sense of humour, and an army pension and the minimum of trapping supported him and his wife. He was interested in everything, particularly in gardening, and, unusually for a trapper, in flowers as well as vegetables. At this time, he was experimenting with wheat and had grown some with only a few inches of straw, but with what object, if any, I do not remember.

All his experiments were not equally harmless. I recollect with pain a new method of curing moose meat which he refused to condemn until its consequences were unmistakable, and an experiment in doping rhubarb wine with the leaves and berries of some plant which, he said, was of high repute among the Indians. A pint or two of this brew gave me one of the worst headaches I ever had. I suspect the Indians esteemed these berries as poison for their enemies. But even if he had poisoned one, it would have been impossible to bear Charlie any ill-will, and he, on his part, had one amiable characteristic seldom found in man, and still more seldom in woman—he never spoke an ill word of anybody. The limit of his criticism was to keep silent.

Charlie and Madge Jones in front of their house at the mouth of the Carbon River.

Shorty Kierce and Tom Stott at Charlie Jones' home.

Toulie on left and Girlie Beattie on right at Charlie Jones' home, ca. 1928.

I only saw Charlie in action once, and then with mixed feelings. We were standing together on the river bank at Twenty Mile when an old, waterlogged boat, which had been washed away in high water from somewhere upstream, drifted past. I thought the lumber might be useful, and put out in a small boat fitted with a 3 hp outboard to salvage it. I had not sufficient power to take the boat in tow, but I hoped to be able to edge it in to the bank as it drifted down stream. Even this proved impossible. The drag of the current was too strong for the engine, and both boats, lashed together, were swept into a back-channel and towards a log-jam.

One of the most awkward things that can happen on a river is to drift on to a log-jam. Usually the boat turns turtle, and to jump on to a log and save one's skin is all that one can do. When we were a few yards from this jam I started to untie the rope which bound the boats together, and shouted to Charlie to untie his end of it. Charlie was very deaf. He could not hear me, and seemed unable to appreciate the situation. He sat calmly gazing down the boat at me, and, as there seemed nothing else to do, I sat down, switched off the engine, and waited to see what would happen.

"Up at Carbon River" – from left: Elizabeth Beattie, Olive Beattie, Tom Stott, Girlie (Mary) Beattie, Madge Jones, Toulie (Louise) Beattie, Shorty Kierce, Bobby Beattie and Charlie Jones – Photo from Hudson's Hope Museum.

The two boats struck the jam with a nasty crash, and the old one which was nearer the jam, started to turn turtle and sink, taking us with it. Then Charlie came to life and slashed through the rope with his knife; at the same time, I dropped an axe across it at my end. The old boat sank immediately. We eased onto the jam, but for some reason kept an even keel, balanced precariously against a log. I looked over the stern and found that the propeller was in a tangle of branches, so, with all my weight on the inside gunwale to prevent the boat turning over, I put a foot and a hand on the nearest log and edged the boat along to a better position. Charlie sat impassively in the bow. After a while I was able to start the engine, and we drew away from the jam. When we reached land again and were able to converse, all Charlie said was; "Too bad to lose the lumber." It was the unnecessary loss of a good rope that was in my mind, but Charlie changed the current of my thoughts by stooping down and picking a large and peculiarly revolting fungus, red and yellow in colour. This he broke in two, handed half to me, and started to eat the other half. "It's good," he said. At any rate it was not poisonous.

Jim's only other neighbour lived at Branham's Flat. He was a frail old Irishman named Mahaffy, a relative, he told me, of the professor of that ilk. The professor, I believe, was a great believer in Hell. It is said that, in the course of a lurid sermon in which he described in some detail the weeping and wailing and gnashing of teeth, an old lady, who did not appreciate his evident conviction that most of his audience would experience these tribulations in due time, opened her mouth defiantly and showed him her toothless gums. Mahaffy paused, leaned over the pulpit, and said in an awful voice; "Madam! Teeth will be provided." Our Mahaffy took an equally dim view of humanity, but, perhaps, with more cause. He had lost all his money in the unlucky mining venture on Branham's Flat, and he was an old man.

It was in order to salvage what he could that Mahaffy had come to the Peace River and settled down to live at Branham's Flat. When the company's mining lease expired he staked the ground. In winter he trapped a little; in summer he pottered about among the piles of iron piping, nuts, bolts, cables and the general dereliction left by the company when it ceased work. His existence seemed to hang by a thread, but his disposition was so unamiable that he failed to inspire the sympathy his condition merited. He was the exact opposite of Charlie Jones. I never heard him speak well of anyone.

Such were Jim's neighbours, the only permanent residents on the Upper Peace River. They formed no great market for farm produce, but farming, for Jim, was a way of life. In addition to the land he owned, which included half a section at Twelve Mile Creek[28], he had the use of many miles of free grazing on the hillsides, and he ran about forty head of horses and thirty cattle on them. The horses were periodically hired out to big game hunting parties, and prospectors. The cattle, which, unlike the horses, had to be fed in winter, supplied him with milk, beef, and babiche for snow shoes; local trappers bought beef, butter, eggs, and potatoes for their own use.

Life on the farm was almost self-contained and self-supporting but some things had to be bought. Sugar, salt, matches, ammunition, clothes, and hardware could not be grown on the land. It was here that trapping entered into the scheme of things. It provided, at the worst, that minimum of ready cash that lack of which harasses so many farmers in times of depression. Over a period of years it had done much more than that. It had paid for buying and shipping-in by road, rail, and river from Edmonton all kinds of agricultural machinery, including a reaper and binder, and a complete blacksmith's shop to carry out repairs.

A farmer from the country to the east once told me that he and his wife could live on a minimum of a hundred dollars a year, but they must have that in cash for absolutely necessary purchases. Times were bad (1931); it cost him more to deliver cattle at the railhead than he could sell them for; all his neighbours for hundreds of miles round were in like condition, and all they had to sell or trade were exactly similar articles. They could help neither themselves nor each other. This man had left his wife on the farm and was camped on the banks of the Upper Peace River trying to make a hundred dollars net profit by washing gold out of the bars. And he was not alone; all over the gold-bearing rivers of the west were men in similar circumstances at that time.

In the nineteen twenties there had been emigration from Britain to Canada and from Canada into the United States. In the nineteen thirties the tide changed. There was emigration from the States into Canada, where there was a little government relief, and emigration from Canada into Britain where there was a lot more. No one who lived through that time would wish to minimize its evils in any country, but the fact is that the unemployed in Britain were comparatively well looked after.

It might be expected that in bad times the pressure of competition would fall also on the trapper, but this was not so. Trapping was a protected trade, and, although it was protected primarily for reasons of state, the trapper had benefited from the control. The province of British Columbia drew a substantial revenue from a royalty on exported fur, and in the early nineteen twenties this revenue decreased alarmingly. To increase and preserve it the government found it necessary to regulate trapping.

Until that time trapping had been allowed to regulate itself, and so long as it was confined chiefly to Indians who traded their catches with the Hudson's Bay Company, this self-regulation had worked well enough. The Indian conserved fur from sheer idleness, trapping only from necessity when he could get no further credit. The coming of the white trapper altered this. If the Indian was idle, the white man was too industrious. Unlike the Indian who traded his catch for supplies, the white man sold his for dollars, and was interested in getting as many as he could. Just as the farmer in the east worked his land to death then moved west, so the trapper did his best to gut his trap-line and then move north.

It so happened that this was comparatively easy. The two staple fur-bearing animals, beaver and marten, are not difficult to catch, breed comparatively slowly, and, unlike wolves and foxes, live in one place until by natural increase they reach saturation point. The fall in revenue from fur royalties was due to the fact that beaver and marten were being exterminated.

And so, in 1926 the government of British Columbia ordained that before a man could trap he must procure a licence, and, before the Game Department would issue a licence, it demanded a sketch map of his proposed trap-line, taking rivers and the height of land as boundaries. If no one was already trapping in that area it was registered in the applicant's name. Provided he did not fall foul of the game laws he could trap there indefinitely, renewing his licence from year to year. No one else was permitted to encroach on his boundaries, and, with certain reservations, he was allowed to name his successor. This regulation not only enabled the Game Department to keep watch on each individual trapper and the amount of fur he was taking from a known area, but, in effect, it made the trapper a fur farmer, at least in respect to beaver and marten. Of other fur he trapped all he could, for no one could guarantee that, if he spared their lives, wolves, foxes or lynx would remain in his territory. But beaver and marten were as good as a bank balance to him. It paid him to conserve them, and the government's royalties were saved.

Trapline Registration Certificate, 1931.

The trapper, too, was saved by this limitation of his activities, and the ratio between trappers and their "means of production" remained constant in bad times and in good. Moreover, the Game Department disliked small trap-lines of, say, two hundred square miles or less, but had no objection to large ones because its objective was to ensure that the trapper could support himself by catching only the natural yearly increase of beaver and marten on his line. It was, in fact, anxious to guarantee him a "living wage".

In this manner a society had been created which was, in its way, ideal—if one can dignify by the word "society" one man and no women to every five hundred square miles. Everyone was a capitalist; no one was either an employer or employed. The hardship of the life eliminated both idlers and weaklings, and there was little scope for those whose purpose it is to live on the labours of others. Money was only used once a-year when fur was sold and supplies purchased. Every cabin door could be left open, assistance and hospitality were automatic, and, in the absence of money and women, there were few quarrels.

Unfortunately, it was entirely artificial and carried no implications relative to conditions in civilization. Man is, fundamentally, similar to other animals; they seldom fight except over food or females, or to protect their young.

With trout in the rivers and deer, moose, and bear in the woods, with a farm producing eggs, meat, milk and vegetables, and a trap-line to provide the necessary cash, Jim, and his wife and five children, seemed substantially cushioned against adversity. Physically the life was hard and so far from doctors and hospitals, the possibility of illness or accident was of more serious consequence than it is in a town, but among the many millions of the civilized world very few had anything like the security of Twenty Mile.

Nor did the life of these unfortunate millions provide them with any comparable interests. Here, a pig was not merely fattened up for market; it was killed, scraped, butchered, salted, smoked, and eaten. If a building was needed, timber was felled and it was built. Everything possible was either made or repaired on the spot. And nature provided certain little luxuries; raspberries, strawberries, high and low bush cranberries grew wild in great plenty, and blueberries and black currants in smaller quantities. These were not only pleasant to eat, but the black bear, feeding on them in the Fall, grew fat and of an excellent flavour. There is no cooking fat like the fat of a bear which has been feeding on berries, particularly for pastry. The beaver, too, is excellent eating, more like the dark meat of a bird than an animal, and beaver-tail soup a luxury comparable with turtle. As for the kitchen garden, everything that could be grown in England could be grown on the Peace River, and, for the most part, was grown at Twenty Mile.

Henry with Olive and Girlie Beattie at Twenty Mile.

"Twenty Mile, feeding stock" - down in a little meadow.

Fresh fruit alone was missing from the table, except, perhaps, once a year. The reason for this was not merely the great extremes of temperature, ranging from 90° to 100° in summer to 60° to 70° below zero in winter, but to the sudden rise and fall in temperature during the winter months. The Pacific Ocean is only a few hundred miles away, and at intervals throughout the winter a warm wind known as a Chinook blows overland from the sea, and the temperature rises very quickly. I have known the thermometer to rise sixty degrees in six hours, from 40° below to 20° above zero. When this wind ceases to blow, which usually happens after three or four days, the weather grows cold again almost as quickly. These sudden changes are hard enough on a man, the most adaptable of animals; as for the trees, their fibres crack with a sound like a rifle shot as the weather grows cold, and again as it grows warm, and fruit trees seemed unable to stand these frequent false promises of spring.

We had arrived just in time to pick blackberries and black currants of which there was a particularly good crop that year, and this we did with great industry so that we might have jam in the winter, if it proved possible for us to stay in the bush. Jim told us that we should have no difficulty in getting a grubstake brought upstream by the Hudson's Bay boat from the town of Peace River, and he thought we should be able to buy a trap-line on the south bank of the Peace. He also told us that the only good agricultural land, with the exception of Twenty Mile, was eight miles further down-stream where Twelve Mile Creek joined the Peace River. Here there was another river flat upwards of a thousand acres in extent, most of which was not surveyed.

Jim owned three hundred and twenty acres at Twelve Mile, and had a cabin there which he was willing to lend us for the winter. Only one other quarter section (160 acres) was surveyed. Everything hinged on obtaining the trap-line, and to find out about this we should have to visit Hudson's Hope. In the meantime, we inspected Twelve Mile Flat. Formerly it had been covered with a dense growth of poplar, but Jim had burned it over several times during the previous few years so that much of the land was superficially cleared. The soil seemed rich enough; there was much pea-vine and vetch, growing in places to the height of a man.

A few days after we had finished picking berries, Jack Pennington visited Twenty Mile, and we decided to make a fishing trip to the Clearwater River and catch as many trout as we could to salt down for the winter. We started late one afternoon, but a heavy thunderstorm blew up, and we spent the night at Carbon River. I slept well enough in a tent with Johnny Darling, but, Jim, Jack, and Tom were in a cabin with dirt and bark roof which leaked badly. In the middle of the night Jack shouted to Darling, asking if he had a gun anywhere, and Johnny replied that there was a Ross rifle, loaded, on the wall. There followed a shot, a crash of breaking glass, and some picturesque profanity. We went across to see what it was all about.

It appeared that, in addition to considerable quantities of rainwater, a bush rat had entered the cabin, and Jack had been trying to make an end of it, but the only casualty was Tom who had cut his hand on some glass. The stale, sickly odour of the rat was the last straw, and we decided to return to Twenty Mile. As we made our way down to the boat in the rain and semi-darkness Jack swore that there was a body buried in the cabin—he had smelled it. In a way, he was right. Subsequent enquiry proved it to be a barrel of moose meat. Cured by one of Charlie's patent processes, it had gone the way of all flesh.

We spent the next two days rebuilding a bridge on the wagon road, and then set off again for the Clearwater, this time without Jack. We spent the whole of one day—the third of August—fishing there, and returned the following day. Jim fished only at the mouth of the Clearwater. Tom and I also went up both the Clearwater and Point Creek to the canyons. In the Clearwater canyon Tom hooked a ten-pound Dolly Vardon on the fly. He had thirty yards of casting line and sixty yards of light backing with a two pound breaking strain. The fish could not get out of the pool, but, sounding like a whale, it easily took out more than all the casting line, so that it became impossible to put any strain on it. Tom played it, or, rather, waited on its good pleasure, for three quarters of an hour. I made a gaff by lashing a shark hook to a small sapling, and, in the end, we killed it.

Tom up on the Clearwater River with rainbow trout.

Despite the time so spent, the day's catch was two hundred and twenty-six trout, and filled a small barrel. With the exception of a dozen Dolly Vardon, caught on bait, all these were caught on fly with three rods. Most of them ran from a half to three quarters of a pound in weight. The largest was the ten pounder Tom caught in the canyon, the remaining Dolly Vardon weighing from four to six pounds each. A memorable slaughter!

The day after we had returned from our fishing trip we walked down with Jim to Hudson's Hope, a typical western village of about twelve families. Standing high above the river at the foot of the canyon, it is sixty miles west of Fort St. John, which really marked the western edge of the partly settled prairie country. It possessed a post and telegraph office, a Hudson's Bay store, and an opposition establishment. The road into the prairie country to the east was very poor, but the river, navigable now for large boats, was used more than the road.

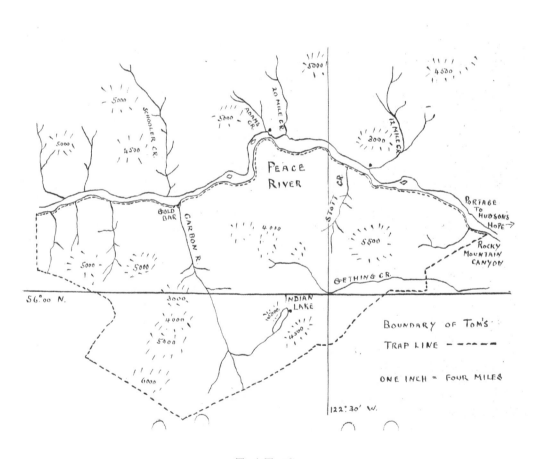

Tom's Trapline

We made several acquaintances at Hudson's Hope, including one with the sinister name of "Deadly" Shaw. We found on enquiry that this nickname was merely a corruption of "Dudley", and even this was not his real name. It seemed that he came from the district of "Dudley Hill", near Bradford in Yorkshire, and had succeeded in giving everyone the impression that the length and steepness of Dudley Hill surpassed anything to be found in the Rockies.

The owner of the trap-line on the south bank of the Peace was willing to sell. Tom bought it, and took steps to have it registered in his name. It ran from the head of the canyon up to Charlie Jones' line, and afterwards Tom bought this also and amalgamated the two of them. The combined lines had a river frontage of fifty miles, eight mountain peaks over 5,000 feet high, and a pretty little lake about three miles long which afforded excellent fishing. The pink, or, rather, red-fleshed Rainbow were its chief attraction, but it also contained a large lake trout said to run up to about forty pounds in weight. These would not rise to the fly, and to catch them one needed about 100 yards of cod-line and a large piece of bait. Tom's technique was to set out in a light canoe with the line fastened to a log, and throw log and line overboard at a suitable place. The log played the fish, and it was only necessary to haul it in when all was over. The only method of cooking these trout was to make fish soup of them. Their fibre was very coarse and oily.

We ordered a nine month's grubstake, and Tom also took possession of the surveyed quarter-section of land at Twelve Mile. Our grubstake contained a number of luxuries, 70 lbs of butter, a case of canned milk and one of canned tomatoes; 10 lbs of coffee; 10 lbs of cocoa and 40 lbs of honey; mustard, pepper, salt, cloves, cinnamon, and various sauces. Of the local fruits of the earth we had our share of the salted trout which filled a five-gallon crock, nine gallons of sauerkraut, onions, potatoes, cabbage, carrots, beetroot, turnips, beef, and pork from Twenty Mile. Tom shot a deer early in November when deer are at their best, and Mrs. Beattie gave us four quarts of pickles and twenty quarts of raspberry and black currant jam. The proximity of Twenty Mile enabled us to enjoy a much more varied diet than that of the average trapper, and far more so than that of the Indian, which is simply meat.

Dieticians tell us that if a man tries to live on steak he will starve to death. The Indian does not try to live on steak; he feeds that to his dogs, and he, himself, eats all the offal: tongue, brains, paws, and, of a moose, the nose. In this way he provides himself with all the necessary mineral salts, and, no doubt, with vitamins also, for his cooking amounts to little more than warming up the meat. He eats no cereals, vegetables or fruit with the exception of the cranberry which, beaten up with bear's grease, is his dessert. This, at least, is his natural diet. Nowadays he is busily trying to live and eat like a white man.

On the eighteenth of August snow fell on the hills, a reminder of the local saying; "There are only two seasons here; Winter and July". This early snow quickly melts, but I have known it to fall in every month of the year, including July. Wintry as this sounds that weather was delightful; cold enough at night to kill off all the flies, and warm, indeed, hot during the day. From the middle of August to the end of October is much the pleasantest time of the year, particularly for horses, which in the winter have icicles hanging from their noses and lips, and in late spring and early summer are eaten by flies.

The middle of August is also harvest time, and, for a while we turned ourselves into farm labourers. We rose at five in the morning, breakfasted off porridge, bacon, and eggs, or steak and chips, and innumerable hot cakes; worked until midday, lunched off meat, vegetables and fruit pies; worked until six o'clock; supped as we had lunched, and put in two hours work before retiring for the night. It seemed a long time since we were living our free and easy camp life, drifting down the river singing sea shanties, and lying out at night watching the stars and the northern lights. I remembered a remark made by a trapper travelling upstream with his squaw and two half-bred children who had spent the night at Point Creek when we were camped there. He praised the varied diet he had enjoyed at Twenty Mile, but added with conviction: "Farming's no life for a white man."

We assisted at cutting and stacking forty-five loads of oats; the remainder was cut with reaper and binder, and when the sheaves had been shock-up, we felt that we deserved another fishing trip. But first we built a smokehouse in order to experiment with the Dolly Vardon trout. We built it on top of a little hillock from the remains of a derelict scow—the only means of getting lumber unless one whip-sawed it. The fireplace, an old oil drum, was dug into the hillock at the bottom, and the smoke piped up through the floor of the smokehouse. This gave us smoke without too much heat. For fuel we used the small, green shoots of the willow.

Before we left the mail arrived. This was a monthly event, by courtesy of the mail carrier, between May and September, when it was carried by water from the Portage to the Hudson's Bay Company's post at Fort Graham on the Finlay River. At Christmas first class mail only was carried by dog team over the river ice, and in March all the mail by horse and sleigh. The mail carrier, Ole Johnson, who trapped on the Wicked River, was a very sick man. He had developed toxic goiter[29], the only "local" disease, neglected it until he was almost too weak to move, spent ten days in hospital for an operation, and was back on the river before the scar had healed. "What is his nationality?" I asked Jim. "He's a Viking," said Jim, "but he'll kill himself as sure as God made little apples."

We left for our fishing trip with Jim and "Shorty" Webber, who was returning to his trap-line up the Finlay River, and camped the first night at Schooler Creek where Shorty gave us for dessert blueberries he had gathered on the banks of the Ingenika. He had a very stylish and well-found grub box and it was said that he never strayed very far away from it since an occasion on which he had nearly starved to death. A band of Indians who owed Shorty money had suddenly broken camp and departed in a suspiciously unobtrusive manner, Shorty, bent on debt collecting, packed a little food on his dog, Kaiser, and followed them. He travelled for a few days without overtaking them, and his food gave out. He travelled a few days more, living on the scraps the Indians left behind but still failed to catch them. There was no game in the country, and, finally, he killed Kaiser, and ate him. In the end he caught the Indians and all was well, except with Kaiser of whom he had been very fond. So it might well be that Shorty liked to know where the next meal was coming from, and I sometimes fancied that "I am so lonely thinking of you," the only distinguishable words of his only song, might have an unusual significance.

We spent the next morning fishing, and caught about a hundred trout; then we returned to Twenty Mile. The river was low, and navigation difficult. Running the Ne Parle Pas the propeller struck a rock, cut the shear pin, and we had to use paddles. Fortunately, we had just made the turn under the reef, and we came through without trouble.

Back at Twenty Mile we smoked the Dolly Vardon trout after leaving them in salt for a day or two. The small willow shoots gave them an excellent flavour and were, as we later discovered, equally good for bacon or ham. Mrs. Beattie salted down the rest of the catch while we stacked the sheaves and took a boat load down to the Portage for winter use there. As the Hudson's Bay company's boat containing our grubstake was not yet due at Hudson's Hope, we returned to Twenty Mile and spent a few days digging potatoes. On the fourteenth of September, while so employed, the first aeroplane to be seen in this part of Canada passed overhead. It crashed further up river in the smoke of a forest fire.

As the Hudson's Bay boat was now due, we returned to Hudson's Hope, spending the night at the Portage cabin on the way down. A cabin here was formerly known as "Cust's House", so named after Bill Cust, a miner who came into the country in the early eighteen sixties. A number of miners came into the country at that time, some of whom had started in the Californian gold rush in forty-nine, then moved north to the Cariboo country, and, finally to the Peace and the tributaries of the Finlay River. Some of them were still remembered, including Bill Cust, Pete Toy, and Twelve Foot Davis. Pete Toy, a Cornishman, famous for his strength and good nature, had a bar named after him on the Finlay River. He was said to have washed a small fortune in gold out of it, cached it, and then perished in the Black Canyon of the Omenica River. Twelve Foot Davis was only of average height. He had received his nickname while prospecting in the Cariboo country. The permissible length of a claim was one hundred feet. Davis had measured the two richest claims and found them to be two hundred and twelve feet. He staked the odd twelve feet and was said to have washed $15,000 out of it.

We crossed the Portage in one of Jim's wagons driven by "Old Ed", ex-engine driver and goodness knows what besides. The brake on the wagon was worked by rope and pulley, and had no locking device, so that a brake-man was quite useful. Some of the gradients were bad, and in places the surface consisted of a clay known as "gumbo" which became as slippery as ice on a wet day. Ed told us that on his last crossing, as he pulled in the horses at the top of a gradient and wagon and horses, as usual, commenced to slide down it, his brake-man lost his nerve, released the rope, and jumped off the wagon and into a wooded ravine. Ed was left to grope for the rope and control the team as best he could. These complications did not, however, prevent him from expressing his opinion. "God damn you," he shouted to his disappearing companion, "I hope it kills you."

We crossed without trouble, and settled down in Jim's house to await the arrival of the boat. Fred Chapman, ex-British Navy, ex London Fire Brigade, was running a small restaurant in the village at this time, and occasionally we dined out. At this season of the year a number of travellers pass through Hudson's Hope; prospectors returning from a summer's work, trappers going in for the winter, and the like. There was also "The Judge" who made a yearly circuit through the Peace River country. I suppose it would be called a "circuit", but a great deal of it was a fishing trip, and for this purpose he always came downstream from Summit Lake. He carried one of Hardy's "Allinone" rods, and was, I fancy, much harder on the trout that on any criminals who came his way, unless, perhaps, they were guilty of stealing food, which in certain circumstances in this country might almost amount to manslaughter. The judge had graduated in Alaska.

With him, this year, came Van Dyk, Inspector "D" Division of the game department of Northern British Columbia. He must, I think, have been a lineal descendent of William the Conqueror who is said to have "loved the tall deer as if he were their father", and he usually made his trip through the country about a month later when there was more chance of catching someone anticipating the trapping season. This lasts from the first day of November to the last day of February, apart from a special late season for beaver. He had complete powers, for in addition to being a sergeant of police and an inspector in the Game Department, he was also a magistrate. He was a very good fellow, when the administration of the game laws was not in question.

Owing to the difficulty and expense of administering justice in the more isolated parts of the country there were at this time, four sergeants of police in British Columbia who were also magistrates. They were expected to avoid trying their own cases, but this was not always possible. One of them, stationed at a very isolated post, was obliged to try minor cases himself, and, a few years after he had taken up his duties there, I enquired how he was making out. "Fine," replied my informant. "Last spring he arrested a man for stealing, prosecuted him, tried him, and acquitted him for lack of evidence." That he was able to do this as a warning to the man, solemnly report it to headquarters, and receive the approval of his superiors (it was the Commissioner, himself, who informed me), instead of a black mark for bringing a case on insufficient evidence, is an interesting side-light on police methods in the bush. After all, the whole business cost the state nothing more than the man's keep for a few days.

While we waited for the boat, Tom made a trip across the Portage with a wagon and team, and brought over a mining outfit. It was in the charge of a man named Featherstone-haugh, a hardy old Northumbrian who boasted that, for the greater part of his life, he had consumed a bottle of whisky a day, and a pound of tobacco a fortnight. It seemed to have agreed with him, for he was in his seventy-seventh year, and lived two years longer.

Meanwhile, I remained at Hudson's Hope, sitting most of the day on the bank of the Peace River in the warm September sun. Day after day was fine and cloudless, but the extremes of temperature characteristic of the climate were now beginning to appear. Already, on the eighth of September, the thermometer had touched 90° in the shade during the day and had fallen to freezing point at night. The cool air, slowly warming in the morning sun, was most exhilarating, but the afternoons, hot and quiet, passed like a dream. The Peace River glittered below its high northern bank, and the poplars, of which there are many hereabouts, had turned a lovely golden brown. Occasionally a canoe crossed the river bringing an Indian from Moberly Lake on a visit to the Hudson's Bay store, and now and then one heard the sound of horses' hooves on the hard, dry ground. Very slowly the sun sank towards the Rockies, and the shadows lengthened imperceptibly as one looked out over the river and the vast prairie lands towards Athabasca. An undeveloped country, yet it had an interesting history, for it was the scene of the final round of the great fight between the Hudson's Bay Company and North West Company for the fur trade of Canada. A fight which nearly ruined both of them and led to their amalgamation in 1821.

It was plain to see that it had a future. The country was immensely rich. Before me lay the Peace River Block, thirty-five million acres of farming country, said to be capable of producing more wheat than all the rest of Canada produced at that time—the last farming frontier. East and West to the north lay the great northern forest belt, the finest fur preserve in the world. There are immense supplies of timber and potential hydroelectric power. Already, before 1916 Lord Rhondda[30] had staked coal claims at Carbon River and drilled, without success, for oil near Fort Vermilion. There was copper, lead, silver, gold, mercury, tin and tungsten in unknown quantities, and the pitchblende deposits near Great Bear Lake were second only to the Belgian Congo as a source of radium at that time.

As long ago as 1874 Captain Butler had selected the Peace Pass as the best route for a railroad through the Rockies—better either than the Yellowhead or the Kicking-Horse Pass. At that time these had the advantage of being further south, nearer to the settled districts and the port of Vancouver. It seemed that the future might well lie with the Peace, and the port of Prince Rupert. But it seemed a very distant future. A railway could not be built until the prospect of freight was sufficiently attractive, and there was enough wheat and minerals in more accessible places. There seemed only two possibilities if events were to move more quickly, a government-sponsored scheme of settlement in the Peace River Block, or a really large gold strike in the Rockies. Neither happened; there was a twist to the future of which I did not dream.

I had come here to trap, and I was more interested in our own small affairs than the future of the country. Next year we should have to build a number of small trapping cabins on our line, about ten feet by five, preferably with a "scoop" roof. This type of roof, only possible over a small span, is made by cutting poplar logs of adequate length and scooping out the inside, and must be laid over a couple of inches of earth to conserve heat. We planned our headquarters cabin on the bank of the Peace River on more ambitious lines. One needed space for comfort, a floor that could be kept clean, and a roof made of "shakes"—hand made shingles split from the first few cuts of a straight-grained tree. An axe and a cross-cut saw would be almost our only tools, and a straight inside wall could only be made by hewing off the round of the logs before building commenced. There are several types of corner-round notch, notch and saddle (a "V" shaped corner), or, best of all, the dove-tail.

We should need a cellar, and this would have to be at least four feet underground to keep out the frost. We should have to build a cache, and this should be about eight feet above ground, the supporting posts tinned to prevent mice climbing up them and in-set under the floor so that a bear could not climb up and break in the door. I spent some time making a list of things that we should need to ship-in from Edmonton—cooking stoves, heating stoves, windows complete in frames, and various tools.

In addition to building there was transport to consider; overland by pack horse, if we cut a suitable trail, and on river by boat and outboard motor, if nothing better could be devised. Horse transport I left to Tom, but I spent a long, time thinking about boats and engines. By this time, I knew the qualities desirable in a boat on this river. As for engines, there was nothing at this time to compete with the outboard, although it had disadvantages. The strain of running twelve or fourteen hours at a stretch was hard on both man and engine, for these engines were built for speed on dead water. They tended to achieve higher power-for-weight ratio as design improved, and to become more fragile and less dependable for our purposes. But they possessed several decided advantages over an inboard engine. On a swiftly flowing stream a propeller must turn quickly to catch the water, and they had a high rate of rpm. Steering was positive, by propeller, not by rudder, and this was of supreme advantage in a rapid. The propeller was not keyed onto the shaft, but held in place by a shear-pin which sheared if one struck a rock - a frequent occurrence - and minimized damage to the propeller. There was nothing more suitable than the outboard at that time.

My dreaming and list-making came to an end with the arrival of the Hudson's Bay boat and our grub-stake. This we freighted over the portage by road, and then upstream to Twelve Mile Creek. The cabin there was comfortable and weather-proof, but it stood in a clump of spruce, and we felt somewhat hemmed-in. However, we did not expect to spend much time in it, and we settled down contentedly enough to prepare for the winter.

———

An ideal trap-line is a circle about eighty miles in circumference with a small cabin every ten miles, and a head-quarters to which the trapper returns every eight days. The headquarters cabin should be on the banks of a navigable river, or, at least accessible by pack horse. Here the trapper keeps most of his supplies, and here he spends two days in every ten, washing, baking, stretching fur and resting. Then the round begins again, and in this way he visits every trap at least once in ten days. If two men are trapping together there should be two such circles, intersecting at the headquarters cabin where they will arrange to meet. If one fails to turn up on time the other can go to his assistance, and prevent what may have been a minor accident from turning into a fatality.

The reason why eighty miles is a desirable length for a line is that a trapper likes to find his catch alive when he visits a trap. He can then kill and skin it on the spot when skinning is easiest, and there is less weight to carry. If the animal dies in the trap the chances are that some other animal will come along and tear up the skin. In any case it will be frozen solid, and must be thawed out before it can be skinned. It is unlikely that any of the more intelligent animals will be caught within three or four days of a visit to a trap, and it will probably live three or four days after it is caught. As ten miles a day, with some sixty sets of traps to attend to, is about as much as a man can cover in the short winter days, eighty miles is about the right length for a line.

The trapper's routine is, however, dependent on the weather. The cold is not excessive in this part of Canada; about 30° below zero is an average winter's day, and this is a pleasant temperature for travelling. But the Chinook winds—a warm breeze blowing in from the Pacific—cause great and sudden variations, and are troublesome to the trapper. He has to carry clothing suitable either for very cold or comparatively warm weather. Rubber overshoes, for instance, made on a moccasin last, for moccasins of Indian tanned moose-hide, excellent in cold weather, stretch and become spongy when it is damp. If the snow becomes sticky and clings to the babiche of the snowshoe instead of sifting through the mesh, travelling becomes well-nigh impossible. A snowshoe is about as long as the height of the wearer, and projects a foot or two in front of his toes. If any weight of snow builds up on the point of the shoe, the strain on the muscles and tendons of the ankle which have to lift it at every step becomes intolerable, and cramp of the Achilles tendon (mal de raquette) is likely to set in. In Chinook weather one travels only to get back to one's headquarters cabin, as there will not be sufficient food at an outer cabin to last for a prolonged stay. As soon as the weather tightens up again, one has to go over the whole of the trap-line, springing and resetting every trap, for a drop of moisture on the trigger will put a trap out of action if it freezes.

In very cold weather, which means about 50° or more below zero, one does not travel either, except to get back to headquarters. Fur-bearing animals themselves hole-up in cold weather, and there is no point in visiting traps. There is no pleasure in it, either. Breathing is slightly difficult, for the moisture on the small hairs inside the nostrils freezes, and one has to breathe partly through the mouth. The cold air burns one's throat, and when it is colder than 60° below, any sudden exertion which necessitates quick breathing for long, may result in frozen lungs, and this is fatal. A trapper does not keep a thermometer at every cabin, and cannot always tell the temperature very accurately, but, at about 25° below one notices on the cabin floor a band of white frost a few inches wide where cold air comes in under the door. This frost forms no matter how hot it is inside the cabin, and it broadens as the weather grows colder. I once asked Jim how one could tell when it was too cold to travel, thinking that a trapper might have some rule of thumb to guide him. "Stand at the door of your cabin," said Jim, "and take a deep breath. If you fall down and can't get up again it's too cold to travel".

A trapper does not wear much clothing however cold the weather may be. He has to keep warm, but he relies chiefly on his own exertions to generate heat. He must avoid perspiring, for sweat will freeze on him when he stops to attend to a trap. Very thick wool underwear, an ordinary khaki shirt, bib overalls (without coat), two pairs of thick wool socks, and Indian tanned moose-hide moccasins on his feet, a pair of cowhide mitts on his hands for handling traps, and any kind of hat, or none at all, is adequate until it gets colder than about 15° below. When the temperature falls lower some kind of wool headgear, a pair of wool mitts under the cowhide mitts, a pair of thin wool trousers under the overalls and a light parka are added, and, however cold it gets, that is enough. The Parkas we used to wear were made by Mrs. Beattie out of Grenfell cloth[31]. There is no heat in this fabric, but it is exceedingly light, and an excellent wind-break. The hood laced round the face and the wrists were held tight by elastic so that the heat generated by the body was prevented from escaping.

Clad in this fashion, with a small pack on his back containing extra clothing, particularly socks, moccasins, overshoes, a little food, and a trap or two, a 30.30 Winchester carbine in one hand and an axe in the other, the trapper covers about ten miles a day, for the going is rough, and always up or down hill. He likes to set out in the morning as soon as it is light enough to travel. Some make a practice of carrying a small lantern for the first mile or so. This is less inconvenient than being benighted. He feels cold at first, and warms up slowly. Whenever he stops to attend to a trap he is conscious that the time he can spend without movement is strictly limited. Perhaps he finds a wolf or fox in the trap. If it is alive, he kills and skins it on the spot. The heat of the animal prevents his hands from freezing during the twenty minutes he is at work, but his body rapidly cools off. If the animal is dead it will, of course, be frozen stiff. This is a nuisance, for he has the whole weight of the body to pack, and skinning is never easy as when an animal is freshly killed, but he must pack it to his next cabin, and thaw it out. As he walks along his eyes are mostly on the ground, partly because he is looking for the trail of any fur-bearing animal that may have passed. If the trail he, himself, is following has become obscured, he will raise his head to look for the next blazed tree.

Much of his travelling is up or down the banks of streams, or close to them, for his trail is usually a game trail, and it is in the valleys that these trails run. Often he has to cross and re-cross streams, and, at times it is convenient to travel on the ice. Here he has to take special care, for streams behave queerly in winter. Sometimes after freezing up, the water will start to flow over the ice, and then freeze again, perhaps only thinly.

Sometimes, after good ice has formed, the water underneath will fall or disappear entirely. Two inches of ice resting on air will not support a man. And there are springs which never freeze properly, even in the coldest weather.

I remember once walking over a stretch of river ice which I had been in the habit of crossing for weeks when suddenly, with a crash, I fell through. For a moment I wondered wildly how quickly I could get rid of my snowshoes and pack; then, my feet struck the river bed, and my chin remained level with the ice. All the water had disappeared. I had nothing worse to do than walk towards the bank, and chop my way out with my axe, and I had time to spare a thought for the unfortunate trout; goodness knows what had happened to them.

Both Tom and Fred Chapman have the distinction of having fallen through the ice into the middle of the Peace River, and lived to tell the tale. Luckily, they were not wearing snowshoes at the time, and were within half a mile of a cabin. One cannot continue a journey even with wet feet, and their clothes were frozen solid before they reached the cabin. Frozen feet are not, as a rule so serious as frozen hands, but inability to travel to one's headquarters where one has plenty of food, has been known to prove fatal.

Arriving at the end of his day's journey the trapper lights a fire in the stove (it is an invariable rule to leave a plentiful supply of shavings after each visit), fetches water or chops ice, or even melts snow if ice is not available, and then spends half an hour chopping wood. It is advisable to keep a good supply of firewood in case he should fall sick, but, in any case, he cannot remain in the cabin until it has warmed up, and he cannot stand still outside. It is partly for this reason that he likes to arrive in daylight. It is foolish to use an axe in the dark, when a glancing bit may mean a nasty wound.

His day's catch will depend on the kind of country through which he has been passing. He may have run a special line for marten up some high, wooded valley. This is easy trapping, for the marten, like its relative the weasel, is all courage and no brains, and will walk into an obvious trap to the lure of a dead rabbit or whisky jack, or even for no reason at all. It is a small animal, easy to pack it if should be dead and frozen, and a valuable fur.

The mentality of the weasel (in winter, the ermine) is extraordinary. One winter a weasel lived under our headquarters cabin, and we welcomed it as an efficient mouser. It never became friendly, but sometimes, attracted by the smell of blood, would come into the cabin, and one evening it trotted round, stopped and put its front paws on my leg and looked at me with its gangster eyes, wondering, it seemed, if I were good to eat. At this time, we had traps set under the eaves of the roof as we had been troubled by bush rats. These were marten traps, several times the size of a weasel trap. When a rat was caught the trap fell to the length of its chain, and one of us would go out and shoot the rat. But sometimes it was the weasel that was caught, and always in the same trap. If it had decided to go that way, it went, and nothing would deter it.

Usually the trapper has the chance of a mixed bag; fox or prairie wolf, lynx, marten, fisher, and, perhaps an odd trap set where he has seen otter tracks may be productive. Fox, at least the red, cross, and silver (the white fox runs further north), and wolves are difficult to trap. They are very cunning and suspicious. As a rule, they know very well when there are traps about, and will even avoid a tree which has fresh axe marks on it or a newly broken branch. Wolves and wolverine will often pit their intelligence successfully against that of the trapper, and steal his bait or eat his catch. Sometimes the only way to catch them is to leave an obviously sprung trap with a dead rabbit in it, and then to make a "blind" set some yards away down the trail. Then the wolf, having skirted the trap, or, very likely, eaten the rabbit, may walk unsuspectingly into the hidden trap. In any case the traps must always be well hidden, and the surroundings apparently undisturbed.

Lynx are quite foolish, but cannot be tempted by bait. They prefer to kill their own food. But they are very susceptible to certain scents. Unfortunately for the trapper, the fashion seems to change every season. These scents are compounded by mixing the scent glands of the beaver with catnip, oil of rhodium, and so on, and a little alcohol to prevent freezing. If a lynx passes a trap without any sign of interest, the mixture must be modified; once it is right it will attract any lynx which smells it. This type of scent is not obnoxious to man. The type we used for fox, which was compounded by collecting fish guts in a sealed jar, and leaving it to stand all summer, was terrible stuff, and would contaminate everything in one's pack. But it was very effective.

Beaver trapping is a specialty. It has a different technique and a separate season, for beaver remain prime until June. Beaver live in burrows in the banks of streams and lakes, making a dam, if necessary, to preserve a good and constant depth of water. Water is his only protection, and the entrance to his lodge is always submerged. He is a clever animal, but does not possess quite as much intelligence as is sometimes attributed to him. A colony of beaver in a back channel of the South Fork of the Halfway River[32] once made the mistake of building a dam so near the head of the channel that they cut if off from the main stream and were forced to migrate. Nor can they fell trees in any direction which they may desire.

The beaver lives exclusively on bark, preferably the bark of the willow. Poplar is a second best, and he will even eat the highly resinous bark of the spruce if he must. During the Fall he has collected his winter's food supply—a mass of willow or poplar sticks which he has built up under the water in his dam. He has been frozen in all winter, and has lived on the bark of these sticks. When trapping him at this time of the year, one cuts a hole through the ice some distance upstream from his food pile, and through it one inserts two peeled poplar poles, side by side, with a trap tied lightly to them in such a manner that when the poles are in position, the trap will be some two feet under water. The chain on the trap is fastened firmly to the bottom of the poles. Above the trap and between the poles a freshly cut stick of willow is placed. The drift of the current carries the scent of the willow to the beaver when he comes out for a meal, and by this time, one may imagine, his own supply has somewhat lost its savour. The peeled poles do not interest him as food; he rises to the surface to investigate the willow, and, if the trap is in the right position, places his front paws in it. When caught his first instinct is to dive; the trap, lightly tied to the poles with thread, goes with him easily, and, as a rule, he swims in a circle round the poles as long as the chain will let him. Then, he drowns.

In the spring, when the ice has thawed out, a different method is used. Two heavy traps are placed under water on either side of the spill-way of his dam. The weight is important; they must be light enough to allow him to take them with him when he dives, but so heavy that he will be unable to swim back. The spill-way of his dam is then broken down a few inches. When the beaver see the water level falling at the entrance to his burrow, he comes out to investigate, and, if the trapper is lucky, puts a foot in one of the traps, dives with it, and is drowned.

Whichever method of trapping is used, the beaver has a quick death. This is not a matter of sentiment but of necessity to the trapper. If a beaver can get his head above water he finds little difficulty in twisting off one of his front paws and escaping. The loss of a front paw is of no great consequence to him.

No reference to methods of trapping would be complete, I suppose, without considering the alleged cruelty involved in the practice. The suffering and inflicting of pain is an integral part of human life, much as we object to it. Cruelty I take to mean inflicting pain either unthinkingly or uncaringly, through callousness, or on purpose, and it may be said at once that the trapper is not guilty on either of these counts. Everything he does or does not do is for a purpose. And some trappers are quite sentimental. I remember one man saying to me "Every time I come to a marten trap I hope to God there's nothing in it." He had a peculiar affection for marten. Trapping happens to be the only way a man can make a living in this country.

Pain is suffered, but it is difficult to form any opinion of its intensity. My own view is that the physical suffering is not great. When the jaws of a trap close the blow stuns the nerves and no immediate pain is felt at all. I know this from personal experience. Rising from taking a cross fox out of a trap I once removed my left hand, which was holding the trap open, rather carelessly. I felt the jaws of the trap brush across the ball of my thumb. As I stood admiring the fox I noticed a patch of blood forming in the snow at my feet. I looked at my hand, and found that the trap had torn a piece of flesh out of my thumb, and the artery underneath was standing up like a cat's back. I had felt nothing.

When an animal is caught the first thing that happens is that the limb, from the point where the trap grips it to the extremity, freezes. An animal sometimes bites off his foot and escapes. It is safe to say that this operation is painless. On the other hand, most animals appear to go mad with a mixture of rage and fear when they are caught. This I am sure is because they are caught, not because they are in pain. What the term "mental suffering" may mean when applied to a wolf or a fox I have no idea, nor has anyone else, but it is certain that some trapped animals do feel this kind of pain to the limit of their capacity.

The case against trapping is that it causes pain in some unknown degree, and that fur, however beautiful, is a luxury. The case is not strengthened by the fanciful stories about the methods and effects of trapping which are told by those who wish it to be abolished. Like the early Christians they seem to think that any story can be told as a fact if its results are likely to be "edifying". But I object only to the extravagances of the sentimentalists. The fact remains that trapping causes pain, and is unnecessary. Fur is not necessary, except to the Eskimo[33], as a protection against cold. The man who faces the most constant and extreme cold—far colder than at the Poles—is the peasant in the east of the great land-mass of Russia, and he does not need fur. But he can hardly be described as a pleasing object either to the eye, or to the nose. It is unlikely that we shall see ladies walking down Bond Street[34] in sheepskin and valenki[35].

The winter's catch depends partly on the trapper's skill, partly on his luck, but chiefly on what is known as the "fur cycle". Most fur-bearing animals feed largely on rabbits, and the rabbit, starting from scratch, gradually increases and multiplies until he is so thick on the ground that the trapper has difficulty in catching anything else in his traps. Then they nearly all die off, and cycle starts again. Like the old "trade cycle", the fur cycle averages about eight years, and the Hudson's Bay Company's records, going back for over a hundred years, show that it has remained fairly constant during that period. Its cause is unknown. The best catches are made the year after the rabbits have died off when there are plenty of fur-bearers, little food for them and few rabbits to get in one's traps.

The trapper knows roughly, before he starts the season, what size of catch to expect. But he does not know what its value will be. This depends on other factors such as fashion, and the amount of money available for luxuries. In November 1929, before the slump on the New York Stock Exchange had affected prices, the average price of marten was forty-two dollars a-skin regardless of quality; in April 1930 it was twenty-two dollars; in May fourteen, and late in the year it slowly fell to eight dollars where it remained for a long time. During this period beaver fell from thirty dollars to eight, and other fur suffered similar declines.

These fluctuations are so great it might seem that in bad times the trapper, himself, would starve to death, particularly so in the instance quoted when the fall in fur values was accompanied by a rise in the cost of living. In fact, living for him is so cheap—so long as he does not come out of the bush—that it is only his savings or his pleasures that are affected. Once his capital expenditure has been made he could, at this time, live comfortably on a hundred and fifty dollars a-year. Even with beaver and marten at eight dollars a-skin, five beaver and fourteen marten would keep the wolf from the door. On the other hand, he never grew rich, although in a good year his catch might cover his expenses six or seven times, and in some cases much more than that. Whatever his luck, his credit was always good.

The life of a trapper has its advantages. He is a capitalist who employs no labour except his own; a worker with no boss. Given good health he is economically secure, and his security is guaranteed in British Columbia by the laws which regulate his occupation. But the great attraction of the life is its freedom. Few men find trapping attractive in itself, and many have skills which they could exploit very profitably in civilization. But they cannot tolerate working to rule. To have to be on the job at a set time, whether you are the boss or a labourer, is something they cannot endure. They work far harder, and for longer hours, than any men I know who work with their hands, but they are their own masters, with no kind of responsibility to anyone except themselves. From June to September when they are, as it were, on holiday, many of them go back into the bush and work even harder prospecting for minerals.

For this freedom the price paid is the loneliness which the trapper has to endure. The Indian has the social background of his tribe, and never strays far from it. The white trapper has no tribe, and his nearest neighbour may be a hundred miles away or more. In daytime the antidote to loneliness is provided by the physical hardship of his life. It is at night, in the dead, white world of the northern winter, that the spirit of man feels the need for communion; it cannot rest in the external, but it can still submerge itself in work. If he is not sleeping, the trapper always takes care to keep busy; it is the only way he can keep sane. In a spirit strong enough to stand it this discipline breeds a self-reliance, which in normal times finds little encouragement in civilization, and is accompanied by a sense of achievement which is equally rare.

We were about to taste this life, knowing little of it except by hearsay, and having no practical experience of setting traps. For this reason, we had arranged to do some trapping with Jim during the first winter, trapping Tom's line only to comply with legal requirements. We assisted in freighting in Jim's grubstake, and then arrived at Twenty Mile prepared to start work.

———

The trail through Jim's trapping country did not form the complete circle of an ideal trap-line. It ran up the north bank of the Peace River to Twenty Mile, then north up the creek and over the height of land and down into the valley beyond to the South Fork of the Half-way River, and eastwards down that stream. It should then have turned south, climbing the divide again, and coming back to the Peace River. But, although a line ran from the Peace River to the head of Twelve Mile Creek, crossing the divide had proved impracticable.

The trail to the Half-way was practicable for pack-horses. It was, in fact, part of the old trail over the Laurier Pass to the Yukon which had been used by those prospectors who had made the journey overland from Edmonton in the days of the gold rush. At the end of September, we left Twenty Mile on horseback with two pack-horses loaded with supplies, and reached the cabin on the summit that night—twelve miles of rough trail rising about two thousand feet. Jim called this cabin "The Pines", for it stood in a clump of jack-pine near a spring. It was large for a trapping cabin, and newly built, but it had no cache, and our first job was to build one. When we had built the cache and provisioned it, we went on to the cabin on the South Fork which stood on a river flat surrounded by burnt-over spruce about ten miles north of The Pines. We provisioned that, too, returned to Twenty Mile, and were preparing for a similar trip up Twelve Mile Creek when Jim went down with appendicitis.

Toxic goiter is the only disease which is endemic in this part of the world, but appendicitis is so common that one cannot help thinking that there may be some special reason for its prevalence. The attack was not acute, but even a "grumbling appendix"[36] needs more consideration in the bush than in town. The nearest medical attention was two hundred miles away, and, once the river had frozen, the going would be hard. Freeze-up was usually about the fifth of November, and it might just have been possible to go out, have the operation, and come back again by water. Jim went down to Hudson's Hope undecided what to do. He discovered a case of oranges there and, after eating little else for a week, felt rather better, and decided to return to Twenty Mile.

"The 'Pines' Cabin. First Cabin up Twenty Mile Creek – very palatial for trapping cabin" – about twelve miles from Twenty Mile.

I brought him up from the Portage by boat, and as we drew near the mouth of the creek, we saw three Rocky Mountain goats on the river bank. Jim was weak from lack of food, but he picked up my rifle, a .300 Savage, and crawled to the bow of the boat. At about three hundred yards he opened fire, and hit one of the goats. We followed it upstream for about two miles where it left the river bank and climbed a steep bluff. Jim emptied the magazine but did not score another hit, and half way up the bluff it disappeared on to a small flat which cut into the hillside. I had no more ammunition, so I tied up the boat and walked back to Twenty Mile to get some, leaving Jim to watch the goat. When I returned and found that the goat had not shown itself, I climbed the bluff. It was very steep, and the surface was loose shale. I could not have climbed it carrying a rifle, so I left the gun with Jim. When I reached the flat the goat stood facing me about ten yards away. I threw a heavy rock at it, and hit it square on the forehead. The goat shook its head and charged. I scrambled a few yards down the hill side, and hung on to a small shrub I had marked growing there. Jim fired again and missed. The goat turned and disappeared over the top of the bluff.

We could not leave it, wounded as it was, so I climbed down the bluff to get the rifle, and then to the top again by an easier route. I followed the goat's trail for about half a mile before I caught sight of it. It could still travel as quickly as I could. I did not want to shoot it in the rump as the expanding bullet would have spoiled half the meat. I fired at its hind legs, and broke one of them. The goat left the trail, and, although I could hear it, it took me twenty minutes searching before I found it in a small clearing in the bush. As I approached it sat looking at me, puzzled rather than afraid, with the most marvellous blue eyes I have ever seen. I paused a moment, almost in apology, and then I killed it.

I did not like to kill except for the pot. No doubt that is why I have never become a good shot. Fond as I am of fishing, I would never catch a trout if I did not know that it would be eaten. Years later Tom gave up shooting in favour of sketching. But Jim's instinct was to shoot first and think afterwards, on the general principle that everything was either eatable or vermin.

Jim had only been a few days at Twenty Mile when the pain recurred, and on the seventeenth of October I drove a wagon down to Hudson's Hope with Jim lying in it on a mattress. He was a bad patient. When we where half way to the Portage he saw a deer near the top of a steep cut-bank, and insisted on shooting it. Deer were now in prime condition and, although we did not need it, we could use the meat. So Jim, who distrusted my shooting, climbed out of the wagon and stalked it. He was about three hundred yards away when the deer saw or scented him. It made a bound, and then stood for a moment on the top of the hill, looking back. Jim fired. The bullet broke its spine and it rolled down the hill almost to our feet.

I left Jim at Hudson's Hope to continue his journey alone, and when I returned to Twenty Mile, I found that Tom had been out with Don McDonald to his headquarters cabin at Crying Girl Prairie, a day's journey beyond Jim's cabin on the South Fork, to help him pack in supplies, and to bring back the horses. He had lost four of the horses, two of them hobbled, and thought they might have returned to Twenty Mile. They had not, and Tom had to return to the South Fork. Here he found them, after searching for two days, travelling east.

"Donald McDonald at his headquarters cabin at Crying Girl Prairie on the South Fork of the Halfway River with two lynx."

Travelling with pack-horses is not my idea of a good time. It is usually impossible to picket them at night because the small patches of grassland do not afford sufficient feed. The most one can do is to hobble the ring-leaders, put a bell on one of them, and sleep on the trail with one ear open. About half an hour before dawn they are sure to make for home, unless one happens to be going home, when they will go in some other direction. So one gets up about an hour before daylight, and, having hunted and found them, removes the frozen hobbles, for there may well be twenty or thirty degrees of frost, thawing them out between one's naked hands. Then one heads the horses back to camp, hoping that someone will have made coffee. Packers with big game hunting parties who may have forty to sixty head of horse to look after are sometimes delayed for days rounding up strays. They are a tight-lipped race, like most men who have much to do with horses, and remarkably profane in a country where no one has much respect for the third commandment.

Towards the end of October, we moved into the cabin at Twelve Mile Creek. This journey was our last river trip that year, for shore ice had already formed, and the river was full of drifting ice murmuring quietly as the current swept it along. We made the journey down stream in a freighting canoe loaded to the gunwales and fitted with a three horse power outboard motor. We did not leave Twenty Mile until dusk. Soon the moon rose, and the water, ice, and snow had the romantic unreality of a picture postcard. It was cold, and with so little freeboard, navigation was not easy in the deceptive light, but we landed within a few yards of our usual place. The next day we carried the motor up to the cabin, cleaned it, and filled the cylinders with engine oil. Then we beached the canoe, and began to think about snowshoes.

Tom landing boat near his house at Twelve Mile.

Tom holding antlers at front gate to winter home.

Little snow had fallen as yet, and in the second week of November we took two wagons to Hudson's Hope expecting to find Jim and some freight there. The freight had arrived, but we learned that, owing to some complication, Jim would not be back for at least a fortnight. We returned to Twenty Mile, and on the way my horses bolted. Harnessing up at Twelve Mile in the morning Tom and I had unwittingly exchanged neck-yokes, and the ring on Tom's was slightly larger than on mine. When I pulled the horses in at the top of the first hill they obediently slowed down, but the ring slipped over the catch on the tongue, and the wagon did not stop. When the wagon hit the horses they gave one big bound forward, and I hit the floor of the empty wagon with the back of my head. Luckily, I kept hold of the reins, and got to my knees in time to swing the horses round an "S" bend, taking an inch of bark off a large cottonwood tree in passing. I had no idea what had gone wrong, but as the horses, although not out of control, refused to stop, I turned them into a patch of saskatoon bushes as soon as we reached the bottom of the hill, and they had to stop perforce. It was not until I started to take stock of the damage that I discovered what had happened. The accident might have been worse, for the reach of the wagon was broken, and the running gear was only held together by a logging chain which, luckily, had been slung between the axles. I had to walk the horses back to Twenty Mile.

On the twenty second of November we were preparing to set out for Hudson's Hope again when the eldest girl[37] arrived at Twenty Mile with the news that Jim would not be returning for another fortnight. She had left Hudson's Hope on horseback in the afternoon of the previous day; ridden the fourteen miles across the portage, rested at the cabin until midnight, and then saddled up again and ridden the remaining twenty miles during the night. She was fourteen years old at the time.

On the same day Tom, who had set a few traps on Twelve Mile Creek, made his first catch—a lynx. It was small, but a fine skin. I stood over it with the Winchester while Tom tapped it on the head with a stick and strangled it. Lynx are peculiar. Most animals object strongly to being caught, even though the physical pain they feel may not be great, but the lynx sometimes seems uninterested, and makes no effort at all to escape. He may have put his pad, as large as a man's fist, into a small trap set for weasel; so small that it can only close over one or two of his claws, and so weak that one pull would free him. But he does not make the effort. He sits down placidly and eats snow for a day or two until the trapper arrives. Then he gives one jump, and is gone. One avoids shooting, if possible, in order not to spoil the fur, but a trapper always likes to keep a lynx covered until he is certain that it is securely held.

We had not started to trap the Twenty Mile line for fear of queering Jim's pitch[38], but it was now evident that he would not return in time, or in a condition, to do any trapping that season. After a conference with Mrs. Beattie, we decided to trap both

Twenty Mile and Twelve Mile for all we were worth, which we did not expect to be much. First we had to get the line up Twelve Mile creek in order, and we started on the twenty forth of November. We had never been over the trail before, and had some difficulty finding it, for it is an old Indian trail, and in places, "blind". When a white man crosses a river or any open ground he makes a large blaze on some conspicuous tree where his trail re-enters the bush, so that it is easy to follow. The Indian does not. He does not wish to be followed, and in such places he makes no blaze at all for about a hundred yards after his trail has re-entered the bush. One can easily spend an hour picking it up. We had also to find the traps which has been left hanging on trees, and set them. That was not difficult so long as we could keep on the trail, but our progress was slow, and we had not reached the first cabin when darkness began to fall. Towards evening I went through the river ice and wet my feet, and we were on the point of deciding to sleep out when we came to the cabin. It had not been used for two years, and the bed was occupied by a bush rat's nest. The sickly-sweet smell was overpowering, but it was cold outside. We burned some coffee on the top of the stove as a fumigator, and tried to get some sleep.

We spent three days on the Twelve Mile line, putting the cabin in order, chopping firewood, and setting traps. Then we returned and went over the Twenty Mile line, spending four days there. It was extremely cold, and we had taken no moccasins, only rubber shoes. In later years Tom and I agreed that this seemed to us to be coldest weather we had ever experienced. I do not know how cold it was, but it was sixty-five below zero at Hudson's Hope at that time. When we had returned to Twenty Mile, I left Tom to trap, and walked down to Hudson's Hope to see if there was any news of Jim. I found that Fred Chapman had quarrelled with his landlord and had moved his restaurant into Jim's house. This was very convenient for me, and that night I sat listening to "Hugh the Drover"[39], broadcast, I think, from Los Angeles, and felt almost civilized again.

The Viking lay dying in the next room, and I went in to sit with him for awhile. He stared at me with his half-blind eyes. "Guess I'm all shot to hell," he said, in answer to my enquiries. He had not improved after his goiter operation, and had determined to see a specialist. But first he had gone to Prince George, bought his winter's grubstake, and started to bring it down stream to his cabin on the Wicked River. The water on the Crooked River was low, and he made slow progress. Finally, he was caught in the ice, had to cache his grubstake, and make his way back to Prince George on foot. From there he took the train to Edmonton, but the doctors could do nothing for him. He had come back into the lower Peace River country by train, and then up to Hudson's Hope to die. He had been a great hunter in his time, and had the reputation of having shot more grizzly bear than any man in the country. He died on New Year's Eve.

Jim Beattie's Twelve Mile cabin and cache.

There was no news of Jim I started back the next day before daylight and walked to Twenty Mile, collecting on the way a fine red fox fur which Tom had left in the cache at Twelve Mile. I took this up to Twenty Mile with me to stretch. A few days later, when Tom and I were at Twelve Mile together, a family of timber wolves came into the country. We heard them howling during the night, and tracked them the following morning. They were running two deer, and had driven them out of the bush onto the ice of a back-channel a mile above Twelve Mile Creek. A few inches of fresh snow had fallen during the night, and the last stages of the chase were plainly written in it. As the deer began to tire they had stopped from time to time, pivoted on their hind legs, and struck at the wolves with their fore feet. But the wolves had kept on, one on each side and two behind, and had finally hamstrung, killed and eaten the deer. A large meal, but the timber wolf is a large animal. The biggest I have seen measured eight feet three inches from nose to tail. Then the wolves had played ball with the heads of the deer which were lying thirty yards away from the carcasses.

Christmas at Twenty Mile, 1930 – in front of the old house from left: Elizabeth Beattie, Toulie Beattie, Clarisse Beattie, Tom Stott, Girlie Beattie, Bobby Beattie, Bob Yeomans, Ernie Gus, Olive Beattie, Fred Chapman, Shorty Kierce, Fred Cassie.

We set some traps, expecting the wolves to return to the kill when they had slept off their gorge, but without much hope of catching one. It is difficult to camouflage traps on wind-swept ice. We need not have taken the trouble. They were asleep on a small island in the river, and when they came back to pick the bones they struck our trail a hundred yards before they came to the kill. They came no further, but made a wide circle, and kept going upriver. They were well fed, and were taking no chances.

On December the fourteenth Tom caught a beautiful cross fox on the Twenty Mile line. A few days later I decided to take a sleigh down to Hudson's Hope. There was a chance that Jim might have arrived by this time; in any case I thought that I would bring the children at school there home for Christmas. They had not expected to spend their holiday at home, and were wild with delight when I arrived, but the news was that Jim would not be back for some time.

Christmas at Twenty Mile, 1931 – in front of the old house from left: Fred Chapman, Olive Beattie, Elizabeth Beattie, Toulie Beattie, Tom Stott, Clarisse Beattie, Girlie Beattie, Bobby Beattie, Bob Yeomans, Jim Beattie, holding a staff and wearing a beaver hat, Bill Kruger, Shorty Kierce, Fred Cassie, and Mac Sturgeon.

When we reached Twenty Mile we found that Don McDonald had come in from Crying Girl Prairie. Tom did not arrive until the twenty fourth, and then we started to celebrate. I had brought a turkey with me from Hudson's Hope, and a whole pig (frozen), and we had received a Christmas pudding and a Stilton cheese from England. So we had the traditional Christmas fare. Then we put on the gramophone and danced until three in the morning, Donald giving us an exhibition of the Highland Fling.

I spent most of my time working the gramophone, dancing not being among my accomplishments, and for some reason, of all the tunes I played that night "O dem golden slippers" is the one that always takes my thoughts back to Twenty Mile whenever I hear it--to the warmth of the cabin and the warmth of the welcome we always received there. It was a happy party, and we kept it going until the old year ended, but in the background was the worry of Jim's continued absence, and the uncertainty of its cause. One could not help admiring the stoicism of Mrs. Beattie; she showed no sign of the fears that must have been preying on her mind.

Tom 'skinning fisher on South Fork divide' - Graham River.

On the second of January Don McDonald and I left Twenty Mile, travelling north. Snow had fallen during Christmas week, and when we had climbed the wind-swept side hills and entered the spruce we had to use snowshoes. We had made a late start, and, perhaps, the Christmas festivities had left us rather out of condition, certainly trapping did not delay us, for there was nothing in the traps save an occasional rabbit or whisky jack, but darkness had already fallen before we reached the Pines cabin. The last half mile of the trail wound through small willows, and we had to strike matches to find our way.

———

In the morning it was much colder. We crossed the two miles of meadow on the summit, and started down the trail to the South Fork. Still no fur. Then, at last, we found a fisher—a small, dark female, the most valuable of its kind. Donald knocked her on the head and skinned her as quickly as he could, for the cold had become intense as the day wore on. We were both glad to reach the small cabin at the South Fork, and to drink a pint of hot, sweet cocoa. I am not a devotee of cocoa in ordinary circumstances, but in these conditions, I know nothing to equal it. Hot rum is excellent when all work is finished, but it makes one sleepy. While there is still wood to chop and a meal to cook there is nothing better than cocoa.

The descent from the summit to the South Fork is step and heavily wooded, but a mile or so from the river the valley opens out, and the cabin stands on a river flat near a ford. Two years later Tom and I spent a very pleasant fortnight here trapping beaver in the early spring. We cut a hole through the river ice, and used poplar as bait. Every morning we walked about a mile to inspect the trap, and, as a rule, took a beaver out of it. On our return to the cabin we skinned the beaver and barbecued it, made soup from the tail, and spent the afternoon scraping the skin and stretching it on a frame made of willow poles. The beaver is a very fatty animal, and it is necessary to clean the skin thoroughly. It is excellent eating, and is the only animal one traps that can be eaten with pleasure except the lynx. Par-boiled, roasted and eaten cold, lynx is not unlike turkey, if you can forget that it is a cat. Although snow was on the ground, and the river still ran under four feet of ice, the days were warm and sunny, both weather and scenery not unlike a fine Easter in the English Lake District.

But now it was mid-winter, and the prospect cheerless. I was not tempted to linger there, and, as Tom had arranged to drive the children down to Hudson's Hope on the fifth, I decided to try to make the journey back in one day, and show him the fisher before he left Twenty Mile. Donald was not averse from an early start. We left the cabin as soon as it was light enough to see the trail, and then parted, for Donald was going further north to his trap-line, and three months of complete solitude.

This was the first time I had used snowshoes, and I had already taken most of the skin off my toes. I found the shoes easy enough to use until I had to climb over windfalls in them, (they were five and half feet long), but I did not know if I should find it easy to cover twenty-two miles of mountainous country in them in a day. However, loafing was out of the question, for the temperature must have been under 40° below zero, and there was no need to bother about traps. I swung steadily along all morning, and it was midday when I reached the meadow on the summit. I found to my surprise that with a clear sky overhead, and the sun beating down on the snow, I should have been more comfortable in sunglasses while crossing even that small, treeless expanse.

On reaching the Pines I brewed tea, and, still feeling energetic, started the descent to the Peace River. The day gradually grew dull and warmer, and when I reached the hills above Twenty Mile a Chinook wind was blowing up. The following morning the thermometer was scarcely above freezing point. It was one of those sudden changes of temperature that cause the trapper so much extra work in this part of the country. Tom was delighted with the fisher. It was the first either of us had seen. It is the big brother of the marten and for some reason is little known in England. The best skins are almost as fine as sable, and much harder-wearing; they are sold chiefly in Paris and New York.

When Tom arrived at Hudson's Hope he found that Jack Thomas, a friend of Jim's who lived there, had left with a sleigh some days before to bring Jim home, and he waited for him. Jim arrived at Twenty Mile on the tenth of January. He travelled over two hundred miles in the depth of winter, mostly by sleigh, and there was not much life left in him. He had gone into hospital a fit man, apart from his appendix, but the operation had been followed by inflammation of the bladder, water on the knee of his right leg, and, finally, a painful sore below the knee. Jim had his own idea of the cause of these complications, and felt that he would not live much longer if he remained in hospital. So he had come home, and, after a few days at Twenty Mile, he recovered sufficiently to express his opinion of the hospital in language that was "frequent and painful and free."

There was nothing we could do except continue trapping, and for the remainder of the season Tom trapped Twenty Mile and I trapped Twelve Mile. I say "trapped", although the Twelve Mile line, after yielding a lynx and a few marten early in the season, produced little else except one very fine cross fox. Tom did very well, and we met about every ten days, either at Twelve Mile or at Twenty Mile, which I visited from time to time. Jim continued to recover, but he was confined to bed, and his right leg was out of action. The swelling on the bone below the knee became very painful. One day he sliced it open with a jack-knife, and drained it. The pain ceased, but the wound did not heal.

First Cabin up Twelve Mile Creek; Henry carrying pack
"The pack was more evident at the time than it is here."

Tom with large Lynx, February 1932.

Furs on Cache at Twelve Mile Winter home – martin, fisher and lynx.

Throughout these winter months we suffered no hardship. Our life was strenuous, but it was not typical of the life of the average trapper. The farm at Twenty Mile provided us with a more varied diet than falls to his lot, and a social intercourse which he never enjoys. The country we trapped was small for two men. We were learning the trade by easy stages, and it is a trade that will bear learning that way. There may be harder lives than that of the trapper. Don McDonald, the tall, lean Islander who came from Stornoway[40] (he objected to being classed as a Scotsman), always maintained that herring fishing was a tougher proposition, and he should have known. But there is no life so lonely; lonely amid surroundings that bear no trace of man's handiwork, and in a climate that constantly reminds him that, if he loses his ability to generate heat, he will quickly freeze to death. In summer there is not this feeling of living on the edge of annihilation; there is always running water, a breath of wind in the trees, the cry of birds and the rustling of small animals in the grass to give companionship. In winter nature wears a different mask, neither friendly nor hostile, but quite alien to man. In the coldest weather there is neither sound nor motion, not the breaking of a twig, the murmur of a stream, nor even a breeze. One lives in a dead world.

Don McDonald's trap-line started at Crying Girl Prairie on the South Fork of the Halfway River three days' journey north of Twenty Mile. The trail, which led to the Laurier Pass and the Yukon, was suitable for pack horses, and at the beginning of October, Donald would pack in his supplies and then bring the horses back to winter at Twenty Mile. About the middle of October he went in himself, alone. Usually he visited Twenty Mile at Christmas, but, with the exception of these few days, he saw no living soul, white man or Indian, from the middle of October to the middle of March or later. The Indian does not trap in this fashion; he does not care even to hunt alone. But this kind of life is the rule rather than the exception among white trappers, modified only when a man has a partner. Then the two meet and spend a day or two together once every ten days.

While making the round of the trap-line we lived simply, and I noticed with surprise that we not only ate less, but seemed to need less in winter than in summer. Bread, butter, bacon, and marmalade at about seven o'clock in the morning (porridge is a delusion and a snare); bread, butter, sometimes potatoes, meat, and jam at about half past five in the afternoon—these were our only meals. In the intervals between trips we experimented in cooking, had a more varied diet, and ate more. My infrequent letters at that time dwelt more extensively on cooking and domestic arrangements than on any other aspect of our lives, particularly when writing to female relatives who found it difficult to believe that a man could penetrate these feminine mysteries.

There was, in fact, very little to write about, but I remember that I had a good opinion of myself as a pastry-cook which owed much to the virtues of bear fat, and also as a compounder of stews. One stew in particular took two days to make, and contained almost everything eatable we possessed. Tom specialized in cakes, and I remember one which contained, among other things, half a pound of salt pork, and a cup of New Orleans molasses. This was a powerful cake. As the Scotsman said of snuff, "Mon, ye ken it's there." But we did not neglect altogether more exotic dishes. As soon as we were settled in our cabin we sent out for a copy of "Mrs. Beeton"[41]. Many of her recipes required ingredients which we did not possess, but we used to read them aloud to each other when we were in the mood, and one Christmas I succeeded in producing with her assistance my chef d'oeuvre—a turkey stewed in oysters.

Our recreations were few. Tom carved a set of chess men out of poplar wood, and we played occasionally, but usually we spent our leisure in reading or sleeping, and I do not remember our efforts at more elaborate cookery with so much enjoyment as two very plain meals which we had a year or two later.

It was in the Fall of the year when Jack Pennington and his partner Fred Cassie swamped their boat in the Ne Parle Pas Rapid, and lost the whole of their outfit and their winter's grubstake. They were lucky to escape with their lives. When they had obtained another grubstake Jack arranged with me to see them through the rapid. I was taking Fred Chapman up to the Ottertail River where he was trapping at that time, and we travelled so far together in two boats. It was the first week in November, and so cold that we had to build a fire and thaw out the gears before we could start the engine, and spray from the river turned into ice as soon as it struck the boat. We lunched at the Ottertail cabin, and Fred came up with us to the Clearwater to lend a hand in the rapid. He brought with him a kitten which, incidentally, behaved in a most surprising manner. When we left the Ottertail this kitten had jumped into my boat, and had become interested in a quarter of moose meat which was lying in the bow. As we drew near the rapid it looked round for its master, and saw him in the other boat about thirty yards ahead. Immediately it jumped into the river and tried to swim to him. The water was rough, the kitten disappeared underneath the boat, and I switched off the engine to avoid hitting it with the propeller. Then I restarted the engine, made a circle, and came up behind it, still battling bravely with the current. I grabbed it by the neck as it drifted past the stern, and threw it on board. It was the sorriest-looking kitten I have ever seen. We signalled the other boat, and handed it over to Fred who wrapped it in his coat and sat in his shirt sleeves in the freezing cold nursing it until we reached the rapid.

Daylight had almost gone when we reached the Clearwater, and when we had finished unloading, we all went up to the cabin to spend the night, sleeping like sardines on the floor. Jack had a half frozen pig in his grubstake, and our evening meal was rum and fried pork. On our return journey next day, the weather was even colder. We lunched with Fred at the Ottertail cabin, and he gave us rum and beef steak. I suppose both the pork and the beef were tough—the rum was excellent—but I never noticed it at the time. I remember only two of the most satisfactory meals I have ever eaten.

As yet we had no wireless, and our reading was restricted. This was, perhaps, in keeping with the life we were leading. Fred Chapman once told me that when he first started to trap he had an old-timer for a partner who took a very dim view of reading. They built one cabin only, as the custom was in those days (a cabin every ten miles is a refinement of civilization), arranged to meet there on a certain day, and went their several ways. Fred was the first to arrive back at the cabin, and when his partner came in, he was reading a magazine. The old man looked at him in disgust. "You're one of these guys who reads, are you?" he remarked, and the next day he started to build himself a cabin of his own. He would not live with a sissy who indulged in reading.

We were not trying to live up to this exacting standard, but in practice we read very little because we had little to read. Each cabin, however, was stocked with at least one old copy of "Punch", and I found myself looking forward, not to opening a new copy, but to rereading one that I knew almost by heart. I can still remember some of the verses in those old copies of "Punch", and the beginning of one gorgeous prose sentence, alleged to be a translation from the Greek of some mythical classic: "Not but what, indeed, on the other hand, O men of Athens ...)".

Trapping, cooking, washing, reading, and sleeping—in this routine our first winter passed, remarkable, as I remember, for one thing only: a curious walking feat, performed by a mad Swede. This man appeared in the country about the middle of February when the weather might have been expected to be most severe. He wore a mackinaw coat, no parka, no snowshoes, and carried no blankets. His mitts were nearly worn out; he wore rubber overshoes without moccasins, and, most remarkable of all, he carried no axe. He avoided nearly every trapper's cabin on his way up river—although I saw him at Twelve Mile—sleeping out at night with what little fire he could keep going with the aid of a knife, and living chiefly on rice. Although one cold spell would have finished him, he made his way a considerable distance up the Finlay River, for, by a miracle, the temperature remained at about 10° below zero for several weeks. Then he developed scurvy, and was cured by a band of Indians who fed him raw moose meat. In the spring he came out, and ended his travels in an asylum. Many men have perished on far less hazardous undertakings than this aimless winter ramble.

The trapping season, with the exception of the spring season for beaver, ended on the twenty eighth of February, and by the middle of March the weather had become much warmer. By day the sun melted the surface of the snow, and travelling became difficult but at night the snow froze again with a hard crust over which one could pass even where there was no broken trail. On moonlit nights travelling was easy and pleasant.

Early in the month we spent a few days at Twenty Mile, killing pigs and smoking the hams in the smoke-house we had built the previous year. Don McDonald was expected about the middle of the month. He did not arrive and we made a trip out to Crying Girl Prairie to see what had become of him. We found him just finishing chopping and stacking next winter's firewood, and travelled back to Twenty Mile together.

In April the low-lying snow disappeared, and the days were warm enough to sun-bathe, although the nights were still cold. The river still ran under four feet of ice, but the small tributary streams had thawed out, and, fed by the melting snow, were pouring an increasing volume of water into the Peace River. At last, on the sixth of May, the pressure became too great, and with a roar the ice burst. Large blocks were thrown up on the banks and stayed there until the sun melted them, but the river carried most of the ice away to be ground to pieces in the Rocky Mountain Canyon. Then spring came in earnest, the woods became alive again, and deer wandered down from the mountains and fed, as placidly and un-frightened as cows, within a few yards of the cabin.

The birds, too, returned. During the winter we had seen none, with the exception of whisky jacks[42], and one snow-white Arctic Owl. In early spring, while the river was still frozen, over a hundred wild swans arrived, and rested for a few days on a small stretch of open water in a back-channel. Later we had many parti-coloured visitors, especially at Twenty Mile, and golden eagles circled again round the crags above Twelve Mile Creek.

With the coming of spring the days lengthened rapidly, but they were not long enough for all we had to do. Jim was still confined to bed, the sore still unhealed, and the leg gradually wasting away. But the farm, which had been lying dormant all Winter now awoke to life. Jack Pennington came in from his trap-line, and he, Don McDonald and the elder children were busy ploughing, harrowing, mending fences and machinery, and rounding up the horses which had been foraging for themselves all winter on the wind-swept side hills. It was a holiday for the trappers, but the beginning of hard work for Mrs. Beattie and the children. Donald, Tom, and I spent a week building a veranda onto Jim's house, so that he could get some sunshine and fresh air; a week of lovely, long days, fine and warm, with the pleasant smell of freshly cut spruce in the air. Then we went down to Twelve Mile where we had much work of our own to do.

Tom's quarter section lay on the west bank of the creek, and the shell of an old cabin which Jim had built years ago was still standing on a hillside about a quarter of a mile from the Peace River. It had no roof, no floor, windows or doors, but the log walls were still sound, and we decided to use them. In addition to reconditioning the cabin, we planned to dig into the bank behind it, and build a cellar store there, connecting it with the cabin by means of a passage; to build a cache to hold such of our supplies as would stand both heat and cold, and to build a veranda onto the front of the cabin.

We spent the first weeks of spring digging out the bank. The ground was still frozen hard to a depth of three or four feet, and we had to cut it out in solid chunks with a grub-hoe. Digging is a brutal kind of labour, and a week or so at a time is enough for any amateur. When it palled on us, we borrowed a horse from Twenty Mile, and felled and hauled the timber for the cellar and cache. While the peeled logs were drying and losing a little weight, we turned our attention to the roof.

We decided to roof the cabin and cache with shakes, the best and most permanent type of roof we could make—better, in fact, than shingles. To be of any use for making shakes a tree must be perfectly straight-grained, and two feet or more through at the butt, and we discovered a suitable clump of yellow pine about thirty miles up the Peace River, just below the Ne Parle Pas Rapid, on the south bank. Early in May when the ice had disappeared, and before the rising water had filled with drift, we set out in our small boat on a shake-making expedition. We lived in the cabin at the mouth of the Ottertail River, and crossed the Peace to work every day. We found that the maximum possible length for a shake was twenty-two inches, and three cuts from the butt exhausted the usable part of a tree. It seemed a pity that so much good timber should be wasted, but we had no use for the remaining wood. After sawing off a round twenty-two inches long, we split it into quarters with wedges, and then cut off the shakes—slices about ¼" x 22"—with an instrument consisting of a heavy iron blade with a handle set at right-angles to it (froe[43]). This blade is held on the log, and struck with a mallet until the shake splits off. We made our mallets from birch which grew close by, and used a great number of them, for the wood was tough.

In five days we made two thousand shakes, and loaded about half of them into the boat. We fastened the remainder onto a raft, and, pushing the raft in front of the boat, started off for Twelve Mile. Our three hp outboard was almost powerless against the drag of the current on the heavy-laden boat and raft, but we managed with some difficulty to edge across the river to the back channel which by-passes the little Ne Parle Pas, and it was not until we had passed Carbon River that we found ourselves in difficulties.

A mile or two below Carbon River there was an ugly stretch of water on the northern side of the river, a sand bar in the middle, and a shallow channel near the south bank. Although we started to edge over towards the channel when we were still a mile away, we could not make it, and we were standing by to cast off the raft, and hope for the best, when it grounded on the sand bar, just above the bad water. There was nothing for it but to cut loose, leave the raft, take our boat town to Twelve Mile, and pray that the river would not rise during the night. In fact, the water fell and inch or so, and the next day we salvaged our remaining shakes, although with some difficulty. They had not been fastened together in small lots, but in one large bundle. When we loosened the ropes, they started to drift down stream.

We had another stroke of good fortune. On the way down river we noticed an old scow stranded on a river bar. It had been carried away from somewhere up river, and, as we knew that no one would attempt to take it back, we requisitioned it. While Tom nailed-on the shakes, I took a crow-bar and a claw hammer, wrecked the scow, and brought the lumber down to Twelve Mile. It furnished us with all the lumber we needed for flooring and shelves.

Lumber was unobtainable locally, and there were other things which we were not fortunate enough to find on sand bars. We had to send out to Edmonton for doors, stoves, windows complete in frames, and, as a luxury, some strips of tongued and grooved oak to cover the old lumber on the cabin floor. I took a childish pleasure in polishing this floor during the winter months when we were able to keep it clean.

We were in no hurry, and this was fortunate, for, in addition to our work, we had to cook, bake, wash, and take time off for fishing. We did no hunting. Moose, deer, and bear are all very poor in the spring, with the exception of a bear which has just come out of his den, and has not had time to work off his last year's fat. Trout, and a little bacon we had left from our winter's supply, provided our only variation from a purely vegetarian diet. Fishing is impossible in the Peace until the middle of July; there is too much sediment in the river, but each pool on Twelve Mile Creek contained trout, and one by one we exhausted those within easy reach.

It was while fishing one of these pools that Tom demonstrated how little a trout minds a hook or two in the tough gristle which forms his lip. He was standing at the foot of a pool, casting up stream, when he hooked a rainbow. The trout dived, took a turn round a sunken snag, and broke the cast. Then it returned to its place in the pool, and continued to feed. Tom tied on another fly, cast and hooked it a second time, but again the trout reached the snag and broke away. This time it came down to the foot of the pool and lay quiescent for a while. Tom tied on a third fly and waited. Soon the trout returned to its place at the head of the pool, and started to feed again. Then Tom crossed to the other side of the stream, and managed to catch his fish. It had the first two flies in its lip.

Occasional visits to Twenty Mile, long talks with Jim, who was still bedridden, and the more varied diet of the farm broke the monotony of our labours. During one of these visits Shorty Webber arrived on his way to Prince George via Hudson's Hope, and Edmonton. It was his custom to make this trip every year, returning with his grubstake down the Crooked River and up the Finlay to the Ingenika where he trapped, and he always spent a night at Hudson's Hope playing poker.

Shorty was built like a bear and had the reputation of being a tough customer. This year he had been living up to his reputation, and it was rumoured that he might miss his card game. Early in the winter he had quarrelled with a neighbouring trapper who, in the course of the feud had shot Shorty's dog. He might, as he said, have mistaken it for a wolf, but Shorty disbelieved this explanation, and had paid him a visit to square the account. Failing to find him at home, he had wreaked his vengeance on the cabin and its contents, and finished off by writing on the door what he proposed to do to the owner when he met him.

Shorty, no doubt, expected this to provoke a personal encounter, but his opponent, impervious to western tradition, thought otherwise. He perceived that the gods had delivered Shorty into his hands, for Shorty had signed his threat. He sawed out the evidence, made a three-hundred-mile trip to Prince George, and presented it to the police. The police did not think it necessary to make a similar trip to arrest Shorty, but it was rumoured that they were waiting for him, no one knew where. Nor did anyone know much about the rights and wrongs of the affair. But public opinion generally was unfavourable to Shorty's opponent who had damned himself in any event by calling in the police.

Shorty, of course, knew what to expect and had no intention of running away. Meanwhile, the forces of law and order were in a difficulty. They felt that Shorty should be arrested at Hudson's Hope, his first contact with civilization, but they had no policeman available. In the end, Fred Monteith, the postmaster, was appointed a special constable for the occasion.

Hating the job, but seeing no way of avoiding it, Fred determined that Shorty should, at any rate, enjoy his customary game of poker, and it was not until the party was adjourning for breakfast that he approached his victim. He cleared his throat and hesitated; Shorty looked up at him and grinned. "It's all right, Fred," he said, "I'll go quietly." So Shorty proceeded at the government's expense to Prince George, where he was going in any event, and where, to the general surprise, he escaped with a substantial fine.

We had intended to make a yearly trip up stream to Prince George, but this year we were too busy, and we ordered our next year's grubstake from the town of Peace River, and a number of items of equipment from Edmonton. Grubstakes for Jim, Don McDonald, and Charlie Jones were also coming from the same place, and when they started to arrive at Hudson's Hope we spent several weeks hauling them across the

Portage by wagon, and then up river by boat. Supplies included petrol, oil and two new outboard motors; altogether we must have handled upwards of fifteen tons of freight.

I found time to bring a mining outfit to Branham's Flat. It was an amateur affair. The members of the party had heard of the gold there, but did not know that Mahaffy was in possession, in a country were possession is rather more than nine points of the law.

I shall never forget the journey. It is about twenty-five miles from the Portage to Branham's Flat, and, normally I could have made it in a day, and been home again before nightfall. I drifted down to the Portage one evening, intending on the return trip to use a new engine which was awaiting me there. When I came to examine the engine, I found that it was fitted with a two bladed racing propeller with an eighteen-inch pitch, instead of the three bladed 12" x12" propeller which we used. There were five of us in the boat, and about two tons of freight, and this propeller would have been unsatisfactory if we had been on dead water. I had a spare engine with me which I carried more as a collection of spare parts than for use, for it had a loose, leaky petrol (gas) tank, and very little power, but, as I found it impossible to change over the propeller, I was obliged to use it.

After making slow progress up stream for some miles, I stopped to refill the tank, and, on starting up again, the engine back-fired. The flash ignited the mixture of oil and petrol which had accumulated on top of the cylinders as a result of the leaking tank. The flames licked up towards the tank, but, fortunately, it did not explode. It seemed a very long ten seconds to me while I unscrewed the engine and dropped it overboard. It took me a great deal longer to fish it out again, dismantle it to get the water out of the cylinders and carburettor, and re-assemble it. We had only reached Twelve Mile at midnight, and I took the party up to the cabin there, baked bannock, and put them up for the night. We made an early start, and reached Branham's Flat in the afternoon.

These prospectors came to some arrangement with Mahaffy, and worked for a time on the Flat. They had little satisfaction, and, later in the year, the mail carrier gave them a lift down river, and dropped them on another gold-bearing bar on the south bank. Here I found them in the Fall, almost out of food, and unable to leave the country without first crossing the river. They were preparing to build a raft when I passed one evening, and saved them the trouble by freighting them across.

This wasted effort was on a comparatively small scale, but it is typical of the country. Neither brains nor brawn, hard work nor enterprise give any assurance of success. Of the many failures I remember two, in particular, which deserved a better fate.

One spring four men drove up to Summit Lake in two old, dilapidated Ford cars. They had a supply of food, a few tools and a little money. With the money they purchased lumber, and built a scow. Then they dismantled one of the Ford engines, converted it into a steam engine, and fixed it in the scow. They added a boiler made from a fifty-gallon petrol drum, and a paddle wheel, cut some firewood and disappeared across the lake in a cloud of steam. They did all this six hundred miles from the nearest machine shop.

There is known to be gold where the Nation River joins the Parsnip, but the gravel is large, the water is swift, and, up to this time, no attempt to work it had succeeded. These men had an idea for a suitable type of dredge, and on arriving at the mouth of the Nation River they constructed it, using the steam engine as their source of power. But they did not succeed in getting the gold, and, as they were unable to return up stream with their outfit, they sold it to a local trapper for what it would fetch, and made their way out on foot.

The second expedition took place the same year. It was an effort of brute force and ignorance. Six Swedes, who had a little money in hand, decided to spend a year trapping. They visited the Game Department, consulted its maps, and found a large area in the heart of the fur-bearing country where no one was trapping. It was situated about a hundred miles up the Finlay River, and then a hundred miles to the north, and difficult to access. But the difficulties which they thought had deterred others did not deter these men. They outfitted themselves for a year, bought two boats at Summit Lake, and accompanied by the wife of one and the sister of another, came down stream to Finlay Forks. Here, the boats separated; one, which was fitted with an outboard, proceeded fully loaded up the Finlay to the point where it was intended to leave the river. Here a cache was built and the provisions stored.

There are no horses in this part of the country, and these prospective trappers needed horses to relay their stores a hundred miles overland to the north. It was for this reason that the boats separated, and the larger, manned by four men, turned east at Finlay Forks, and came down the Peace River to Twenty Mile where Jim sold them four horses. Their boat was forty-four feet long, fitted with a twenty-five horse power marine engine, with a speed of 800 rpm. They had plenty of horse power, but the propeller did not revolve fast enough to catch the water on a swiftly flowing river. New to the work, they had not realized this on the journey down stream; when they turned to go upstream they found that the engine could not even hold the boat against the current. And so, for eighty miles and through two rapids, these four men had to manhandle their boat and horses.

After some weeks they reached Finlay Forks, met their companions, transshipped the horses into the other boat, and, late in September, reached their trapping ground. Then they discovered why no one was trapping there; a forest fire some years earlier had gutted the country. The married man, his wife, and a friend continued north regardless of trapping boundaries until they reached the Muddy River. They built a raft, drifted down the river until freeze-up, and then built a cabin, and stayed there through the winter. And there the woman gave birth to a child. The following year they all came out to Prince George. Their combined catch sold for three hundred dollars.

———

It had been a lovely spring, but by July the year had grown old and dusty, and the flies, although not so bad as in the previous year, were annoying. We began to take life more easily, working less on the cabin which was now weatherproof, and spending more time fishing. About the middle of the month we started to catch large Dolly Vardon trout in the Peace River, and a little later we could catch Arctic and Rainbow on the fly. Long stretches of river seemed almost devoid of trout, but here and there they congregated in large numbers in the eddies, and as soon as the water had cleared they were not difficult to catch. The fishing lacked the interest of an English trout stream where the fish are more sophisticated, and demand a more natural presentation of the fly, but it was pleasant to spend an evening in a canoe drifting idly in an eddy, and catching without effort the next morning's breakfast. Even when dusk had fallen we had no inside attraction, for we had no light except candles and a small oil lamp, and we only entered the cabin to eat, sleep, or seek shelter.

Jim's leg showed no improvement; the sore had penetrated deeply, and the leg itself had shrunk and become stiff. He had born these months of bedridden idleness with a kind of impatient good humour; now he began to think that his leg would need more expert attention than it could get at Twenty Mile. He decided to go to Edmonton where, presumably, adequate medical attention could be obtained. Towards the end of the month I drove him down to Hudson's Hope, and he set out on his five-hundred-mile journey.

I was at Hudson's Hope on the first of August when a Junkers hydroplane, the first aircraft to visit the country, landed on the river. It was attached to an aerial surveying party, sponsored by the government of British Columbia in conjunction with the Canadian Pacific and the Canadian National Railway companies, and was making a survey of the country up to ten miles north of the Peace River. The government was interested because this northern part of the province had as yet never been surveyed, and the existing maps were considerably out of drawing. The two railway companies were planning, at least on paper, to continue the Northern Alberta Railways which was slowly pushing its way from Grande Prairie to Dawson Creek. They surveyed it through to Prince George by the alternative routes of the Pine Pass and the Peace and Parsnip Rivers.

While Jim was away either Tom or I stayed at Twenty Mile until the middle of September when we spent a week exploring Tom's trap-line. The country was new to us, most of the trails were "blind", and, apart from the fact that there was a cabin at the head of a lake somewhere near the foot of a mountain, the peak of which could be seen from Twenty Mile, we knew nothing about the country. We packed oatmeal, flour, baking powder, salt, sugar, chocolate, raisins and tea, but we forgot to take any bacon, and our tea ran out about the middle of the week. We crossed the Peace River, and followed the trail up Fisher Creek, camping the first night on its bank.

It was very hot, and our forty pound packs, axes, and rifles seemed an intolerable burden, but most unexpectedly we found a number of wild raspberry bushes full of ripe fruit. The next day we turned east, and discovered one of the cabins. It was very damp, and the walls were covered with some kind of large fungus inside. It seemed hardly habitable, even in winter. The third day we reached the mountain[44], and, after climbing a long way up it in a vain endeavour to obtain a view of the surrounding country, descended in search of water, and found a lake[45] at the foot of its southern slope.

We came onto the foreshore just as dusk was falling, and a wolf setting out for its night's hunting, howled plaintively across the water. I looked across the lake, and presently a coal black timber wolf came out of the bush and trotted along the foreshore. I judged it to be five or six hundred yards away, and put up the sights of my .300 Savage a notch before taking a shot at him. The wolf started, but trotted unharmed. I shot again with no effect. I felt that I was firing high, and, as the wolf seemed likely to escape, I handed the rifle to Tom who is a much better shot than I am, with a remark to that effect. Tom took a leisurely sight and fired. The wolf spun round, jumped into the air, and disappeared behind a clump of willows. "Put a shot into the willows for luck," I said, and Tom did so. The next day, when we had rounded the lake, we found the wolf, one bullet through its chest and the other through its stomach. It is good shooting to hit a running wolf at upwards of five hundred yards; to hit an invisible wolf at that distance savours of the black arts.

Tom also shot the head off a sitting grouse with his Winchester. The bird was very welcome, for we had eaten no meat or fish since we started, and by this time had run out of tea. We were tired of living on bannock and water. Then, one morning, in a small pool at the head of Carbon River, we found our first trout. They were about six inches long with the typical colouring of English brook trout unlike any we had seen in Canada, and the pool was alive with them. I lit a fire and unshipped the frying-pan while Tom tied a hook on a few feet of line. He dropped the naked hook into the water, and the small trout fell on it like minnows. He pulled them out, one after another, and I dropped them into the frying pan. We ate them, bones and all.

The country through which we were travelling had been rich in beaver, but it had suffered severely in the years before trapping came under control. We found several small streams bordered by acres of meadow land where there had formerly been beaver dams. The dams had been cut, and beaver either exterminated by trapping or driven away and the old workings stood like ancient, ruined cities. There were a few beaver left in the lake itself, but these had eaten all the poplar within easy reach, and were now gnawing spruce. The further they had to go for food, the more vulnerable they became to their enemies, for a beaver is helpless on land. Even if they were not trapped at all their rate of increase seemed likely to be slow.

At this time of year we could form no opinion about the number of marten on the line. There were plenty of wolves and grizzly bear, but neither were of much importance as fur. Fortunately, there were no Indians. Many years ago a camp of Indians had nearly perished while wintering in this country. The grizzlies killed some of them, but their chief trouble had been lack of food. The moose seemed to have deserted the country. The Indians came to the conclusion that there was a curse on it, and never came back.

We discovered the five cabins the line boasted. The cabin at the head of the lake was usable, and the only one which had a proper cache. The others were all very small and in bad condition. The best of them had no door; it was built halfway up the side of a ravine, and one entered through a hole about two feet square high up near the roof. We decided that there was enough building required to last us several years.

When we had completed the survey of our line we returned to Twelve Mile for another kind of survey, for we had decided to purchase most of Twelve Mile flat, and the government surveyor in charge of the railway party met us there and surveyed it for us. We had no ambitions in the direction of agriculture beyond the desire to keep a few horses. A farmer is tied to his land and his stock. One of the chief attractions of our life was that we were free, any time the fancy seized us, to get into a boat and travel for hundreds of miles, and we had no intention of forfeiting this freedom. But it would have been awkward for us if this land were to pass into the possession of anyone else.

The survey completed, Tom went up to Twenty Mile and out to Crying Girl Prairie with Don McDonald who was packing in his winter's grubstake and did not wish to come out again with the horses. I went down to the Portage to pick up a ton or two of freight for Charlie Jones. After making this trip I packed the outboard motors away for the winter, and sent out for a number of spare parts. These engines stood up marvellously, but they were not designed for the type of work to which we put them. The only one in running condition was the old crock with the leaky tank.

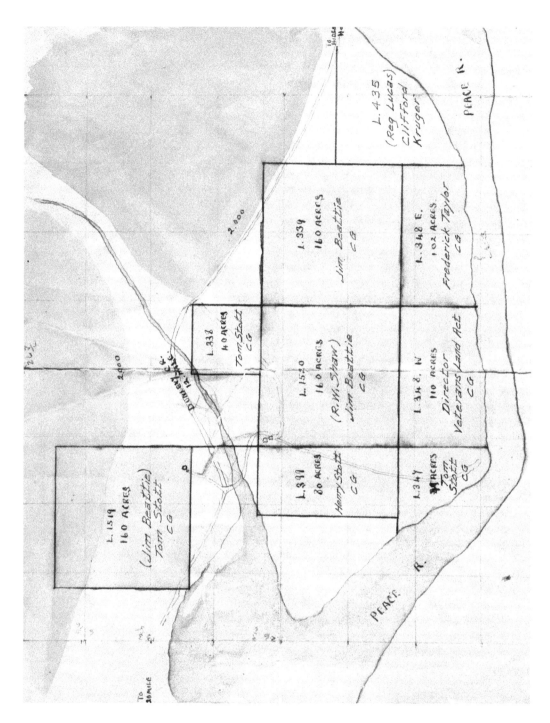

Map made by Tom Stott of some District Lots at Twelve Mile showing expected 2,000 ft. flood line. Names in brackets are of people who requested the land survey but did not receive the Crown Grant.

B.C. Names Card – Stott Creek

All the parties on the railroad survey had now completed their work, and the white triangulation marks on the highest points of the surrounding mountains were the only traces they had left behind. They had christened one of the mountains to the south of Tom's trap-line "Beattie Peaks", and one of the streams "Stott Creek", thus giving us a kind of immortality. But it was not the fate of the Peace River to run alongside a railway. It was eventually built through the Pine Pass. A road also was constructed so that tourists from the States can now make a circular trip from Vancouver to Prince George, on to Fort St John, and back via Edmonton. Many of the men in these parties were trappers who had followed up a winter at their trade with a summer spent in climbing mountains. Naturally, they were thirsty. They had made arrangements to quench their thirst at Hudson's Hope, and from time to time we heard accounts of the way they had painted the town red when they finally reached that settlement.

At this time Hudson's Hope possessed a young policeman who took himself rather seriously, and, as no one wished to fall foul of the law, some of the party locked him in his barracks before proceedings commenced. All unpleasantness was avoided by this thoughtful act, and the party, although rowdy enough, passed off without harm to anyone. There was only one incident which affected the local inhabitants. One of the celebrants, a man named "Skookum" for his size and toughness, wandered out of the village and found himself in the early hours of the morning near a farm house. He entered the veranda, and the farmer who happened to be sleeping there got up, and asked him what he wanted. "How far is it to town?" enquired Skookum. "About two miles," replied the farmer. Skookum got into the bed and pulled up the blankets. "Better beat it to town, son," he said, drowsily, and fell asleep.

One day towards the end of October, Jim's brother arrived at Twenty Mile, very late at night. He told me that he had brought Jim up a far as Twelve Mile in a wagon, and, as Jim found travelling by wagon very painful, he wanted me to fetch him up to Twenty Mile by water. I was doubtful whether the old engine would stand the trip, but, after a few hours sleep, I got it out of the cache, and carried it down to the river. The boat, luckily, had not yet been hauled out of the water. It was a cold morning, the river shrouded in a white mist, and the engine disinclined to start. After a while it burst into a half-hearted roar, and clouds of vapour from the exhaust mingled with the morning mist. We soon reached Twelve Mile and had Jim aboard, and the engine managed to hang together during the return trip, although, as the petrol was gradually consumed, the lightened tank danced about alarmingly. It seemed impossible to brace it effectively with hay wire, and soldering aluminum was beyond our power.

Jim had spent the summer months having the tendons of his leg stretched by means of weights and pulleys. Then the doctors had put the limb in a cast, and told him that he could go home for the winter, although he had explained to them what going home meant. We had already moved into our cabin on Twelve Mile Creek, intending to trap Tom's line during the winter. In view of Jim's condition, we changed our plans, and decided to trap Twenty Mile again, and share the proceeds with Jim. On Tom's return from Crying Girl Prairie we went out on the line.

When we returned to Twenty Mile we found Jim's leg out of the cast. It had become very painful, and finally turned blue. It seemed that the cast must be interfering with the circulation. The knee joint was now quite obviously dislocated. Jim had had enough. He had been in bed for a year, and a trapper cannot lie on his back for ever. The disease was getting no better, and no one seemed to know what it was. I suggested that his best plan would be to cross into the States to the Mayo brother's clinic at Rochester. There, at any rate, he would get the most expert attention in the world. As Jim could not travel alone, I offered to accompany him there, or elsewhere. Jim agreed, and we set out to find a doctor who would either cure the limb or cut it off.

We had no time to lose. The river was already running slush ice, and at any moment the weather might make travelling impossible for him until a sleigh could be used. Early next morning Mrs. Beattie left by wagon for the Portage, and in the afternoon Tom and I set out with Jim in the Peterboro freighting canoe. It was canvas covered, but we were going with the stream, and we handled it delicately amid the ice. We crossed the Portage the next day, and were fortunate enough to find Van Dyk, who had called at Twenty Mile while Tom and I were on the trap-line, at Hudson's Hope. He had a boat and engine, and was going down river to Fort St John. The ice was much thicker on the lower river, but again we were going with the current, now in a stoutly built wooden boat. After a day at Hudson's Hope, Jim and I departed with Van Dyk, leaving Mrs. Beattie and Tom to return to Twenty Mile.

We travelled by car from Fort St John to Pouce Coupe, and then by train to Edmonton. Here Jim was visited by several old friends who had trapped and packed with him twenty years ago when the Grand Trunk was under construction. They gathered round his bed and begged him with tears in their eyes to lead a more sober life in the future. "Damn it," said Jim, "I used to spend half my time putting these guys to bed in the old days." They all looked very fit, although they were living more sedentary, but possibly more sober lives than they had done in the past, and one of them was not long out of jail where he had served time for stuffing ballot boxes with non-existent votes. They were anxious to know what had happened to Jim, and he told them. He had a good vocabulary, but his opinion of the medical profession had passed beyond mere words although he did his best.

As a result of this meeting we changed our plans. One of Jim's friends convinced him that there was a doctor in Prince George who would give him satisfaction, and it was to Prince George that we went. The doctor took a series of x-rays, and sent them for an opinion to the best bone specialist in Canada. The opinion was that, if the disease could be cured, which might take a matter of three years, an operation for ankylosis[46] would then be performed, and Jim would have a stiff leg. But a month later a second series of x-rays showed that the disease was getting worse, and had started to affect the bone above the knee. It seemed likely that, if the leg were not amputated now just above the knee, it would have to be taken off later at the hip.

Once the decision to amputate had been taken things began to move. The operation took place on the eighteenth of December. On the sixth of January Jim was out of hospital, and before the end of the month we were back at Twenty Mile. We travelled via Edmonton and Grande Prairie, where Jack Thomas met us with a sleigh. It was a cold trip up the ice on the Peace River, forty to fifty below zero, and Jack Thomas froze his nose and ears on the journey. But, in spite of everything, Jim began to improve rapidly. There was now no chance that old Charlie Paquette would come up to Twenty Mile to "plant" him.

This was an old, rather grim joke between Jim and Charlie. Years before, Charlie had eaten some bad canned meat, and suffered a severe attack of ptomaine poisoning. He was lying in a tent across the river from Hudson's Hope when one morning Jim appeared with a spade, and announced cheerfully, "You don't seem to be getting any better, Charlie. I've come to plant you." Charlie was furious. He swore that, come what might, he would not die at Hudson's Hope, and he persuaded an Indian to put him on a horse and take him over the hills to the Indian settlement at Moberly Lake. He did not die, perhaps he was too angry, and he did not forget. Whenever he met one of Jim's children while Jim was ill he would ask, "How's your Dad?", adding, "Tell him if he doesn't get well soon. I'm coming up to Twenty Mile to plant him."

So far I had spent the winter eating, drinking, and playing cards. I had been made an honorary member of the Legion Club in Prince George, and had enjoyed much hospitality at the home of the surveyor we had met on the Peace River. The New Year's party was particularly impressive, at least, to me. It started at six in the evening of the 31st of December with a dance and continued until half past seven in the morning on the 2nd of January when a heated argument arose as to whether dawn was breaking or night was falling. An argument which, fortunately, was solved naturally in about half an hour.

Tom had a successful trapping season while I had been growing fat and flabby with idleness and good living. He looked me over disparagingly when I returned, and enquired if I proposed to trap. "I think I'd better," I said, "if only for my health's sake. Show me where the sets are on Twelve Mile, and I'll take that part of the line over." "Yes," said Jim, vindictively, "take him out and kill him. He's given me the hell of a time."

Tom did his best. There was only enough bedding for one at each cabin, and on our trip over the line I had to pack a sleeping robe in addition to the usual impedimenta; rather a heavy pack which I was very ill-fitted to carry. We reached the first cabin with an hour or so of daylight to spare and decided to go on to the second cabin which was only about four miles away, and save a day. The snow was deeper on this part of the line, and after a mile or two of it I had had enough. For the rest of the journey I collapsed in the snow every fifty yards to cool off my face and allow my heart to slow up a little, while Tom waited patiently until I was able to continue. In this fashion we finally reached the cabin, and that evening I shirked my share of wood-chopping and cooking.

The following day we visited the traps at the far end of the line beyond the cabin, and then returned to Twelve Mile.

Perhaps it was the result of my poor physical condition that I set out on my next trip in a very bad temper. I felt so annoyed, quite without reason, that I did not even trouble to follow the trail, but endeavoured to walk myself into a more reasonable frame of mind by taking short cuts through the bush wherever possible. This was heavy going, but all went well until I lost myself, some distance above the first cabin, among a series of beaver dams. This was easy to do, as the trail was only indicated by a bent willow twig here and there. Having lost it, I should, of course, have backtracked until I had found it again, but I had not yet recovered my temper.

I ploughed on through the willows, and the going became more and more difficult. I could make no progress at all among the thick willows with an axe in one hand and a rifle in the other, so I tied my axe on my pack. Surrounded by trees, I could not see the hills, and the day was hazy, with no sun visible. Travelling in a series of twists and turns I had to depend on my sense of direction to take me towards the cabin. After some time, the excessive exertion cramped the muscles of my left leg, and I could no longer lift the point of my snowshoe. The tie worked loose, and I stopped to retie it. A prairie wolf sneaked through the willows about twenty yards away, but it was gone in an instant, and I had leaned my rifle against a tree.

I plodded on. I was not lost in any the real sense of the word, but I began to think that I might have to sleep out that night. Then I discovered that my axe was missing. This was awkward. I could not be sure of finding the trail in the little daylight that was left, and, if I had to spend the night out, I needed the axe. I decided to go back and look for it at the place where I had stopped to retie my snowshoe, and there, by great good luck, I spotted about half an inch of the haft sticking out of the snow. I went forward again, and after a while came to another beaver dam, and on the far side of it I saw a blaze on a small spruce. I was glad to be on the trail again, for by this time my left leg had become very stiff, and scrambling about in the willows had become hard work. My bad temper, however, had quite disappeared.

During this winter I spent only one month trapping, and had little to show for it except the loss of the superfluous flesh I had acquired in Prince George. Early in March Tom and I foregathered at Twenty Mile. Jim, who was now getting about on crutches, wanted a large barn built, and, while the snow was still on the ground and hauling timber was easy, we felled and hauled the trees. Afterwards we spent some days trapping beaver on the South Fork, and when Don McDonald returned from his trap-line, we joined him at Twenty Mile, and together built the barn.

In the course of building I unintentionally provided Jim with some amusement. Standing on top of ten rounds of logs I dug a peevee[47] into the eleventh to hold one end in position, and shouted to Donald, who was handling the horses, to take a pull on the other end to swing it about a foot. I bent my weight on the peevee, the horses pulled, the handle of the peevee snapped, the log fell inside the building, and I danced a hornpipe in an effort to keep my balance. Jim, who had told me I couldn't hold the log against the horses, was highly delighted, although it cost him a peevee, and we had to suspend operations and chop the log out.

We had decided that this year we would go up river to Prince George for our supplies, and then on to Vancouver to see a dentist. The plan suited Jim who had to go to Calgary to be fitted with an artificial limb, and Don McDonald, who had not been out of the bush for some years, offered to accompany him. Jim, however, had to wait until six months from his operation which meant that we should miss the best water. We spent the intervening weeks making improvements to the interior of our cabin and cellar at Twelve Mile, and Tom found time to shoot two black bear before they had worked off their last year's fat. We also burned over the flat at Twelve Mile, as this had not been done for some years.

First we backfired at the eastern end of the flat, and protected the cabin in a similar manner. Then, choosing a day when a steady westerly breeze was blowing, and the undergrowth was just dry enough to burn, we set fire to the western end. The fire burned all day, and far in to the night, covering the whole of the flat and running up the side hills to the top of the first bench, but no further. Although safe enough, for our backfiring had been done efficiently, and, in any case, it was too early in the year for a real forest fire to have started, it was an awe-inspiring spectacle, and I have doubts about its legality.

Towards the end of May, we made preparations for our journey. We re-caulked the boat, and went over the seams with pitch, dismantled the engine and reconditioned it. Both engines and boats needed this attention every spring, for the engines had usually broken down by the end of summer, and boats were made of spruce. There seems to be no limit to the capacity of spruce to shrink and check. When the boat had soaked up, and the engine had been tried out, we baled up our fur, collected provisions for the trip, and set out.

The prospect of retracing our journey of two years ago was full of interest. It was, in a way, a "final act". The journey up river to Prince George was all that remained to round off our experience of the life which we were to lead. This year it had an added attraction for me, because it was a race against time. We were a month late in starting, and had to travel as quickly as we could. Our boat was very lightly loaded, but our speed averaged only about five miles an hour. We covered some sixty miles a day; twelve to fourteen hours on the river, and hour for lunch, and the remaining nine to eleven hours for making and breaking camp, cooking, eating and sleeping. Hard work, and for the most part on strange water.

The river from the Portage to Point Creek had by this time become familiar to me, but from Point Creek to Summit Lake it was practically unknown. One learns little drifting down stream, and I had now to run the engine in from two to three feet of water, and to judge the water at sight. This, in its way, was an enthralling pursuit, although the pleasure it gave me was of a very different kind from that of our journey down stream. Mark Twain, who was himself a river pilot, gives in his book *Life on the Mississippi*, the best and most interesting account I know of the art of river navigation. After describing its technique, he concludes: "Now when I had mastered the language of this water, and had come to know every trifling feature that bordered the great river as familiarly as I knew the letters of the alphabet, I had made a valuable acquisition. But I had lost something too. I had lost something which could never be restored to me while I lived. All the grace, the beauty, the poetry had gone out of the majestic river." This was the difference between our return to Summit Lake and our journey to the Peace River two years ago.

We left Twenty Mile late one afternoon, and spent the night at Carbon River. Charlie was in good form, and regaled us with some excellent rhubarb wine which on this occasion had not been reinforced with "some berries which the Indians use." During the summer he had enjoyed an encounter with a raconteur of his own calibre. A member of a surveying party, after listening to one of Charlie's stories, had gravely related how, on a big game shooting expedition, he had climbed a mountain peak and shot a big-horn (Rocky Mountain sheep), on a neighbouring crag. He remarked that the rarity of the air at great heights aided long-distance shooting, and spoke in high praise of telescopic sights. "How far away was it?" asked Charlie. "I dunno just how far," replied his guest, "but the meat had spoiled before we could get to it."

We left Carbon River early the next morning, and, after passing Schooler Creek and the Ottertail River came to the Ne Parle Pas Rapid. We ran up through the big eddy below and tried to run the rapid, but found it impossible. Every time we climbed over the first big wave the propeller came out of the water, and we drifted back. So I backed slowly out and put Tom ashore with a line. The extra manpower made all the difference at the critical moments, and we climbed slowly through the rough water into the swift, smooth stream above. We drew in close to the shore, and Tom coiled the line and threw it on board. We could not stop to take him on board also until we were well clear of the rapid, for one never knows just when an outboard will refuse to start. As we rounded the point above, a belated piece of drift in the form of a large spruce tree swung round the bend, and passing within a few yards of the boat, plunged into the rapid.

We stopped at Point Creek, remembering the large Arctic trout there. Half an hour's fishing sufficed to provide us with dinner and next morning's breakfast, and we continued slowly up stream until, late in the evening, we reached the Wicked River, and there we spent the night. On this occasion we could not spare the time to visit the canyon, and so, with this ambition still unfulfilled, we made an early start and reached the Finlay rapid after about three hours running. We crept up through the rough water close to the south bank of the river, passed through a back-eddy, and started to climb, but, again, the engine alone proved insufficient. We dropped back in to the eddy, and Tom went ashore with line. There the river bank is rocky and precipitous. Tom climbed a crag, rounded it, and passed out of sight. While we were waiting for him to pay out the hundred yards of line, I gazed across the rapid and noted with awe where we had run it two years ago.

When we pushed off from the bank the engine failed to start. We drifted in the eddy for few moments, and then the main stream caught the bow and flung us out in to the river. For an instant the line grew taut, and I saw Tom standing on the cliff edge, straining on it. If he could have held us we should certainly have swamped, but the rope tore through his hands, and we shot stern first down the stream. As soon as I was assured that Tom was not following us, I straightened the boat with a paddle and turned my attention to the engine. It was soon going again, and we worked our way back in to the eddy. Tom was waiting for us on the bank. When the last yard or two of line had whipped though his hands it had taken most of the skin with it, but he was able to make a second, and, this time, successful attempt.

At Finlay Forks we did not turn south up the Parsnip River, but continued up the Finlay for ten miles to visit Alan McKinnon who trapped there. In an effort to save time we followed the slack water of a back channel for about two miles only to find that a log jam had formed at the head of it, cutting it off from the river. It was well past midday when we arrived at McKinnon's place. We spent the night there, and some trappers from the Finlay joined us; Miller, whom we had met at Point Creek when we first came into the country, and several others. The next day we all left together for Summit Lake.

The journey up the Parsnip became a boat race, and to me the course was unknown. The water was opaque, and, while the main features of a river stand out clearly enough, and broken water is usually easy to read, it is another matter always to keep the boat in from two to three feet of water, and to cross the river at the most advantageous point. Most of the time we took second place, but once we followed a narrow back channel for over a mile and beat our companions into camp by half an hour. A logjam had formed at the head of this channel, too, but there was a gap just large enough to squeeze through. The continuous vibration of the engine had a paralyzing effect on one's left arm, the roar of the open exhaust was deafening, and if there was a following breeze, one sat in a constant haze of blue fumes. Despite these discomforts it was fascinating work, but the beauties of nature passed unnoticed. I can still see stretches of water in my mind's eye, but my memories of the scenery go back to our quiet drift down stream.

The race came to an end when we neared the Pack River, and my engine broke down. The fibre cam inside the fly-wheel which moved the rocker-arm of the magneto had disintegrated. I had with me as spare the old crock with the leaky tank, but I hated to use this, and borrowed an engine from Miller. It was an old, bronze, saltwater model which had once belonged to Jim, very heavy, not very powerful, and rejoicing in the name of "Iron Mike". But it was better than mine, and it stayed the course to Summit Lake. The Pack River was getting low, and we had to pole up Cross Rapid, but in due course we reached McLeod Lake, and passed the night at the Hudson's Bay Company's post there. We sold some fur, spent a restful evening playing cards, and Jim won back Iron Mike.

Travelling across McLeod Lake was a pleasant relaxation, but all too brief. We were soon in the Crooked River, and here the water was unpleasantly low, but still navigable. We crossed Kerry and Red Rock Lakes, and camped for the last time at the entrance to Davie Lake. It was raining, for the first time during our trip, and we made camp, but by four o'clock in the morning the rain had ceased, and we started our last day. We had still about fifty miles to go, and for long stretches we could not use the engine.

Crossing Davie Lake under power in the dull calm of early morning was very different from our frenzied paddling against a gale of wind two years ago, and when we came to those dreary miles of the Crooked River which had then seemed endless, we had to run at half speed to make the turns, and, even then, had soon left them behind. Our trouble came when we reached the riffles. We ran into them at full speed, and when the water shallowed, switched off the engine, tilted it, jumped into the water and grabbed the boat before it had lost way, and manhandled it up stream. These riffles were steep, and the water swift, and at times one ran the engine too long, struck a rock, cut the shear-pin, and drifted down stream before one could check the boat. It was always tricky re-starting the engine in the middle of a swift, shallow run, but it was necessary as soon as the water deepened and the pressure became too great for man power. Then one ran the engine at half speed so that the propeller would draw less water, and Tom and Donald assisted it by poling.

It was a hot day, and the tall spruce on the banks shut off any breeze there might have been. At the head of the Long Riffle, nearly two miles of collar work, we rested and made tea; at the head of the Harrison much shorter, but particularly bad that day, we had a late lunch. As we worked our way nearer to its source the river became more and more narrow, and now, for the most part, we could pole when it was too shallow to run the engine. It was evening when we passed the remains of the beaver dam and at long last came to the slack water and the lily-pads at the very head of the stream.

When we reached the lake we lashed the boats together, set course for the farther shore, and relaxed. It was a lovely evening, and, as it happened to be Sunday, there were a few holiday-makers from Prince George on the water. They seemed like denizens of another planet. Passing a small island, I noticed two; a girl who looked unbelievably ethereal in her light summer frock, and a man. He was a tall, well built fellow, but he looked effete in his white flannels and cleanliness. I glanced round the boats. My companions were burnt a deep brown by the sun and wind, and badly needed shaves and haircuts. Smears of blood and engine oil streaked the tan on my arms. We were all incredibly ragged and wet through to the waist. But, I reflected, there was no office desk awaiting us in the morning.

I filled a pipe, leaned my back against a bale of fur, and, lulled by the hypnotic secondary vibration of the engines, thought of the civilization I was about to re-enter. Hot baths, water in the pipes and beer in bottles; hotel restaurants serving meals twenty-four hours a day. My thoughts centred on these for my desires went no further. How attractive they seemed! No wonder the human race had sacrificed so much to achieve these and similar marvels. But our visions are curiously incomplete. Imagination is the inveterate accomplice of desire; it never pictures the result of fulfilment. Civilized man has escaped from his primeval servitude to nature, but somehow he seems to have lost dignity, and even significance, in the process. And he has not achieved freedom.

The Return

"For nature, heartless, witless nature,
Will neither care nor know
What stranger's feet may find the meadow
Nor ask amid the dews of morning
If they are mine or no."

—A E Housman

MOST OF THE trappers from the north had arrived in Prince George before us, and were now preparing to leave. Among them was Shorty Webber. Shorty was on top of the world. He had been lucky enough to find a sample of ore during the previous summer—his first prospecting venture—which, after being assayed, had received favourable mention in a mining journal. The morning after we arrived he entertained a group of us with a lecture on prospecting, concluding with the remark: "If I had only known as much about prospecting ten years ago as I know now, I should be a rich man today." One of the bystanders, a man known as "Slim", who had prospected in Alaska, Manchuria, Australia and various parts of the American continent, listened gravely, and, when Shorty had finished, summed up briefly: "You must have been some Goddamn son a bitch of an ape ten years ago, Shorty," he said. For a moment Shorty frowned, apparently thinking it out, but the gust of laughter was too much for him, and he was obliged to join in.

For two days we wallowed in hot baths, incidentally, the most exquisite sensuous enjoyment civilization has to offer, then Jim and Don McDonald left for Calgary, and Tom and I for Vancouver. The man who used to drive the stagecoach drove us the sixty miles from Prince George to Quesnel in his taxi. He had a great reputation for handling horses, and I have never seen such hands on a man; his fingers were like a bunch of large carrots. His method of handling a car was to drive furiously for a few miles, then stop and get out with the remark: "There is a spring here; the best water in Canada," upon which I produced the rum bottle, and we sampled the best water in Canada.

When we arrived at Quesnel, we found that the annual rodeo started the following day, and it was impossible to find any accommodation. We all slept in the taxi, and, the following morning, stiff and dazed, we were wandering round Quesnel when we were accosted by two men who in a matter of minutes forced on us a couple of note cases, and some slips of paper, fading away with the parting warning; "Don't take any wooden money, boys." I was somewhat hazy about the nature of this transaction at the time, but discovered later that we had become subscribers to several Vancouver daily papers. For two years afterwards great bundles of these arrived at Hudson's Hope to the extreme disgust of the mail carrier.

We spent a pleasant day watching the rodeo, and then took the train for Squamish. The Pacific Great Eastern Railway was a government venture, designed to connect Vancouver with the Canadian National Railway at Prince George. Squamish is about forty miles from Vancouver, and one has to make the journey by sea at the southern end. The funds ran out a Quesnel, and from there one had to make a similar journey by taxi to Prince George. Apart from this it might be said to have achieved its purpose. It reminded one of the description of the Holy Ghost in the Creed: "Neither born nor begotten but proceeding." But, as the pleasant hilly country round Quesnel and Williams Lake merged into the grandeur of the Coast Range, one was amply repaid for the inconvenience of the trip. It was the most scenic journey I have ever made, and the best approach to Vancouver is by sea.

There is excellent cooking in Vancouver, and when our teeth had been attended to, we proceeded to use them, paying particular attention to Chinese and Mexican dishes as these were as far removed as could be from our usual fare. It took us some time to get used to the traffic, and after a few days we began to get bored with the bustle of city life. It was here, I remember, that I sacrificed my only chance of being converted by Aimèe Semple Macpherson[48] in favour of bathing at English Bay.

Before we left I bought a Colt automatic pistol. The law in British Columbia in regard to firearms was peculiar. One could buy a pistol and keep it in one's home, but it was illegal to carry it without a police permit which was unobtainable even by a trapper who had good reason.

So I was only within the law when this gun was hanging on the cabin wall; taking it there or carrying it afterwards was illegal.

The necessity under which I have found myself from time to time to break the law used to annoy me, but one becomes conditioned to it. I have lived as a member of the English middle class, in a palace, and in the bush, but I have never been able to live anywhere without breaking some law or other. Nowadays in England the law is so extensive and oppressive that nobody knows what it is, except members of the legal profession, and the criminal classes who have to know what they are doing for professional reasons.

When we arrived back at Prince George we found that Jim had acquired two artificial limbs; a peg leg, and a leg with a foot. The Peg leg turned out to be the only one that he could use in the bush. He had also fallen head first down a flight of stairs in Calgary and broken his wrist. This did not delay us, and we left Prince George as soon as we could, for it was now July, and we had two boats and about four tons of freight to get down the Crooked River, where the water was falling rapidly. In fact, we had travelled only a very short distance when we ran aground. The water was incredibly low, not more than three or four inches deep in some places, and it seemed almost impossible that we should ever get the boats down the river. But, by dint of much heaving, relaying the cargo, and at times slewing the boat round, damming the river with it, and floating a few feet at a time on the resulting wave, we gradually worked our way down. We had come up from Davie Lake in one long day; it took us over six days to make the passage back.

The last day was wet and cold and I managed to get myself knocked off the stern of the boat at one of the hairpin bends. When we made camp, and Donald said; "I don't like broaching cargo, but I think we might open the rum," I heartily agreed. I dug the case out, but found to my horror when I came to open it that the liquor store had supplied whisky by mistake. There was nothing wrong with the whisky except that it was 30° under proof, whereas the rum was much more powerful. It was impossible to make a hot drink with the whisky without diluting it beyond recognition. There were loud complaints which the whisky did nothing to assuage, indeed, it seemed to inflame them. I had been responsible for ordering the liquor, and when the conversation turned to various forms of lynching, I promised to get a case of rum in if I had to go out and fetch it.

In the end I succeeded in doing this by writing to the Judge, and asking him to bring one with him when he came in the Fall. I have a suspicion that here again I may have been breaking the peculiar liquor laws of the country, but, if so, I was doing it in good company, and I duly paid a fine. The Judge had his own ideas of law—and equity—and in September he arrived with the rum. "I don't think I've done so badly," he said, with a large grin, "I've only drunk two bottles on the way down."

After shooting the last riffle on the way down the Crooked River, we had no further trouble. Until we reached the Parsnip there were places where the water was too shallow to run the engine but it was always deep enough to float the boat, and we made Twenty Mile in four days. When we had unloaded we settled down to an easy life, at least until harvest time. Fate, having shot its bolt, moved on in search of other game, and in course of time Jim recovered his health. His wrist grew well, and, although he broke a few ribs during an over violent convalescence, he stuck some plaster on his chest and took it in his stride. He would never wear snowshoes and had to find someone to trap for him, but, at soon as he could sit a horse he rode recklessly about the country wherever a horse could go. The loss of half a limb seemed little more than a nuisance to him, and the gruelling experience he had been through seemed to have left his immense vitality unimpaired.

It takes two or three years to become acclimatized to Canada, and in that time we had also learned the essentials of our trade. In particular, I had now acquired sufficient experience of river conditions to design what I thought to be the most suitable boat for these waters. Here I had a stroke of luck, for it was at this time that the Johnson Outboard Motor Company produced an outboard drive for an inboard engine. This drive was very substantially built, and would transmit over 100 hp. It had all the advantages of an outboard motor, excepting that the propeller was keyed-on to the shaft in the ordinary way, but the engine was connected to the unit with a spline coupling which allowed it to tilt if the skeg[49] hit a rock, cutting off the power.

I designed a boat for this outboard drive, and during the winter we had it built at Summit Lake. It was thirty-eight and a half feet long with a fifty gallon petrol (gas) tank in the bow, and a forty hp. racing engine in the stern coupled to one of the first of these drives obtainable in Canada. It was a very satisfactory craft and would run the rapids either up or down in almost any water conditions, but it started badly, and the following year it gave me the hardest river trip I ever made.

The engine was late in delivery, and the following spring I had to stay behind at Summit Lake and bring the boat down river alone. This would have been impossible had not the builder agreed to give me a hand down the Crooked River. All went well until I had run the Finlay and Ne Parle Pas Rapids, and this, at any rate, was very fortunate. But, before I reached the little Ne Parle Pas a grinding noise in the gears and a suspicious lack of power caused me to pull into the bank and investigate. I found that the gears had run dry, and those in the head of the drive were chewed flat. The boat was heavy, having plenty of power I had had it built of thicker lumber than was usual, and I had to continue the journey with paddle and pole. It would have been impossible had I not known the river well and been able to think a mile or two ahead, as it was it involved continuous hard paddling to cross and re-cross the river.

Luckily, the back-channel round the little Ne Parle Pas was still navigable, and I was on the right side of the river to take it. I had a lot of hard work, but no trouble until in changing over from paddle to pole, I carelessly knocked a paddle over board. It was one of a pair of beautiful large paddles which the boat builder had made for me, and I was loath to lose it. It drifted near a steep gravel bank, and, as it was moving more slowly than the boat, I got well ahead of it, and then, carelessly disregarding the swiftness of the stream, worked the boat in near the bank, ran to the bow and jumped with the rope in my hand. I drew the bow of the boat in to the bank, and hung on to the rope. The current swung the stern round; I had no firm foothold on the loose gravel, and the drag of the boat pulled me down the bank until I was up to the waist in water. I determined to go down the river with the boat, if necessary, but the pressure eased, and I scrambled to my feet in time to see the paddle drift past. When I repeated the performance more circumspectly lower down and was ready to capture the paddle, it was sucked down by a whirlpool and never came up.

I continued downstream, but when I drew near to Carbon River I failed to keep close enough to the northern bank, and the boat was seized by a large and powerful eddy from which I was unable to escape. Every time the boat swung round to the head of the eddy, I tried to paddle out into the main stream, but at each attempt the eddy and a north wind held me back. After several unsuccessful efforts I decided to wait until nightfall when the wind might be expected to fall. I tied the boat up, and paid a visit to Charlie Jones. Charlie entertained me royally; towards nightfall the wind dropped, and, fortified by Charlie's home-made wine I paddled out of the eddy and completed my journey to Twenty Mile.

There was more trouble to come. In addition to the wrecked gears, the exhaust flange casting was cracked. I sent out for replacements, but these did not arrive until October, and we lost the use of the boat for that season. In the following March, when I came to examine them I found that the wrong set of gears had been sent, and I had to return these and ask for a replacement. The gears arrived in May, just as the ice went out of the river, and I quickly assembled them. My final difficulty arose in fitting the exhaust flange casting. In dismantling I had to break the union between the two elbow joints in the exhaust pipe which was badly burnt-in—about six inches of 2" iron pipe with a 3/32" thread at each end. Iron piping does not grow in the bush, so I went to see Mahaffy, and was lucky enough to find a suitable piece of pipe which was threaded at one end. I cut off six inches of it, and settled down to cut about an inch of thread at the other end with a hacksaw and a small three cornered file. It took me two days. Then, at last, everything was in running order, and we had no further trouble, and a great deal of pleasure from the boat. When I left Canada Tom did not want to keep it. I sold it to the police, and it became a patrol boat at Fort Simpson on the Mackenzie River.

This was not my lucky year on the river, and in October I had an unusual experience. At the beginning of the month the weather became very cold; slush ice was running in the river, and shore ice had begun to form. By the middle of the month it seemed as if winter was upon us, and when I arrived at Twelve Mile from a trip in the bush I found that one of our boats—a forty-two foot, red painted craft—was lying high and dry on the bank, ten or twelve yards from the falling water, and that thirty yards of shore ice had formed. Twelve Mile was not suitable place to keep a boat during the winter. When the ice broke up in the spring, huge blocks were apt to be thrown onto the bank. I decided to take the boat up to Twenty Mile at once. I spent the rest of that day cutting rollers and levering the boat down to the edge of the ice. The following morning, I chopped a channel through the shore ice, worked the boat into it, lit a fire and thawed out the outboard, and started upstream for Twenty Mile.

There was a lot of drifting ice in the river, and as I worked my way up I found more and more shore ice. At the bend above Twenty Mile the river was frozen completely across. Here, at the point where I wished to land, there was about a hundred yards of shore ice. The edge of the ice was frayed to the thinness of a wafer by the running water, and could not, I thought, be very thick anywhere. I backed the boat across the river, took a run at the ice, and crashed onto it. The ice did not even crack. The boat slid over it, and froze fast. I climbed out of the boat and drove a pole through the ice. It was only three inches thick, but resting solidly on water. It was a long way to the shore, and I decided that the best thing to do would be to go down stream for about a quarter of a mile, and haul the boat out where there was much less shore ice to cross.

I chopped a channel round the boat, and removed as much ice as I could from underneath it. Then I pushed out into the river again and started up. The engine raced, and, looking over the stern, I found that the propeller was only half in the water. A lot of ice must still have been sticking to the bottom of the boat. The water cooling system of these engines depends on the action of the propeller, and if I ran it for any length of time under these conditions it would seize up. I switched off, and tried paddling, but I could not even steer the boat by this means. It seemed that I might drift down river until I reached the canyon.

I started the engine again, and noticed that the propeller gripped the water sufficiently to steer the boat. I switched off; went to the bow of the boat, coiled the rope, and laid it and an iron shod pole ready to hand. When we reached the part of the river where I intended to land I started the engine again, and steered the boat alongside the ice. When the boat was within a foot or two of the ice I switched off, ran to the bow of the boat, picked up the pole and rope, and jumped as far out on to the ice as I could. The ice held; I drove the pole through it, took a turn of the rope round it, lay down and held on.

When the boat had swung round and the drag eased off, I pulled the pole out and fixed it in again as near to the bank as I could. Then I walked up to Twenty Mile, harnessed a team of horses, cut some rollers, and took them and a hundred yards of half inch manila rope down to the boat. The horses were accustomed to ice, and I hauled the boat out without difficulty. But I had worked hard to no purpose. A week later the weather turned warm. In ten days all the ice had gone, and in the middle of November we were still on the river.

———

These troubles, mechanical and other, had to be accepted as incidental to life in the bush; they would have been infuriating in civilization; here they were not so disturbing. And, in general, the life had so far been very satisfying. Trapping, in itself, had never much attraction for me, but the various kinds of constructional work we did, and the hundred and one ways in which we pitted our strength and ingenuity against nature gave a sense of achievement, while the long days on the river in summer, and travelling in the bush in winter, were an endless delight. But gradually, as one mastered the tricks of the trade, and the routine of existence made smaller demands on one's energy, I began to think of other things.

It was not only that at last I had the time for reflection. I felt that, after three years of this life I had sloughed-off my civilized background as far as that was possible, and had become sufficiently familiar with nature to judge what it meant to me. We were surrounded by the most beautiful mountain scenery, but I was seldom conscious of its beauty. We had to carry water a quarter of a mile, and in winter, when the water disappeared, to chop and melt ice, but chores like these were no longer irksome; they had become second nature. I was acclimatized. That first night on the Crooked River I had known that it would be some years before I could answer the question: did the attractions of freedom and natural surroundings outweigh those of civilization? The tiller of the soil lives close to nature, but he has already sacrificed much of his freedom. I approached the question from the point of view of the hunter who lives with wild nature and on it, whose surroundings are "natural" in the fullest sense of the word, and whose freedom is as absolute as man can achieve.

It was not, I reflected, a question of absolutes. Certainly I had freedom, but I had other things also. My chief delight was on the river, and here I had a wooden boat, and a petrol engine; hardly the equipment of a savage. I had also a rifle and a number of steel tools. There was a demand in the cities for fur, and in return they provided me with these things. I was a parasite. But, if I made a dugout canoe with a stone axe, and was content with a paddle and a pole, that would be descending into another kind of servitude which was no part of my ambition. I needed not only freedom, but power. The historic alternative to the petrol engine had been slaves to paddle the canoe, and here one was back in organised society again. The pursuit of freedom was automatically limited by the pursuit of the power to enjoy it, and there were limits to both. The limit, for man, was living in some kind of harmony with society. I was, in fact, making a compromise.

Henry with moose, ca. 1930.

I was making it at the extreme outer-edge of human society, surrounded entirely by nature in the raw. This is a different experience from any that can be enjoyed in the British Isles where nature everywhere bears the marks of man's handiwork. Was this what I wanted? Did nature itself provide me with a satisfaction and an inspiration that human society lacked? For some years I had been buried deeply in learning to live this life, now my mind began to grope for an answer to these questions.

Others, before me, had suggested answers. In the nineteenth century there were many who, in revolt against the growing materialism of the times, the "hubris" with which man pursued his purposes, and the unsightly detritus his activities left behind, had almost worshipped nature. I was on my guard against them, for they had often expressed themselves in language so beautiful as to be compelling in itself, and I knew myself to be susceptible to the beauty of words. I had discovered this when, after reading the Gospel according to St John in the Authorized Version, I had read it in the original Greek. In Greek it left me cold, and I had to turn to Plato's Apologia for any affect similar to that of the Elizabethan English. But I need not have worried. The past three years had been an admirable prophylactic.

I considered the views of these nature worshippers dispassionately. Ruskin, for instance; the concluding paragraph to his "Queen of the Air"; words which have always stuck in my mind: "Ah, masters of modern science, give me back my Athena out of your vials, and seal, if it may be, once more, Asmodeus therein. You have divided the elements, and united them; enslaved them upon the earth, and discerned them in the stars. Teach us, now, but this of them, which is all that man need know, that the Air is given him for his life; and the Rain to his thirst, and for his baptism; and the Fire for warmth; and the Sun for sight; and the Earth for his meat—and his Rest."

Magnificent! But "which is all that man need know?" This static ideal was no part of my philosophy. I felt more in agreement with the man, who, when exhorted by the village parson to thank the Lord for the fruits of a garden he had planted, replied "Aye, but you should have seen it when the Lord had it to himself."

And Wordsworth; brooding upon Grasmere, Wordsworth could write that he had found: "In nature and the language of the sense, The anchor of my purest thoughts, the nurse, The guide, the guardian of my heart, and soul, I Of all my moral being."

Real as the emotion was for Wordsworth, to me it seemed pure hallucination.

Long before Wordsworth the pagan religions were grounded in nature worship, but these old ideas were sunk too deep in a long-forgotten mode of thinking ever to resurface, and nothing that I observed in nature served to rekindle a spark of life in them. I remembered, however, that they were divorced from morality, and in this I felt they were right. Powerfully affected by nature, Wordsworth felt, like them, that so profound an effect must have a spiritual cause, and, like all Europeans after nearly two thousand years of Christianity, he was unable to imagine divinity divorced from morality. It was this that I felt was untrue.

When one thinks of wild nature, one is apt to picture to one's self, mountains, rivers and lakes; prairies, deserts and the sea. But the greater part of nature is cold, empty space. Alone in the bush in winter it needed little effort of imagination to realise this, and to feel, with Pascal "Le silence eternel de ces espaces infinis m'affraie" (The eternal silence of these infinite spaces frightens me). In summer, when one lived the happy life of the extrovert, the alien quality of nature was not so noticeable; but it was there. The seasons pass and return; the rain falls on the just and the unjust; and the random effects of kindness or cruelty are equally purposeless. Nature is full of energy, but it strives after nothing, and its beauty is as haphazard as its ugliness. It always takes the line of least resistance, and, if one can learn anything from this, it is that one should try to produce the maximum effect with the minimum effort, which is, after all, the secret of all effective design. For the rest, all that nature said to me was "sink or swim; eat or be eaten, kill or be killed," and this is a lesson that man has learned so well that he seems likely to destroy himself.

Most men, of necessity, accept their values as they buy their clothes, off the peg. I had come to see if there was any inspiration to be found by living close to nature, and not merely playing at it. I had found that I was prepared to eat anything unless it ate me first. That was all that nature had to tell me. But I had had a magnificent rest-cure. After all, man is himself a part of nature, and the bond uniting them is strong. There are few things in life that give so deep a feeling of contentment as returning to natural surroundings; it is like coming home when the day's work is over, and putting on an old coat. It is not surprising that man has always tended, emotionally, to see nature in his own image, or, intellectually, to see himself merely as a natural phenomenon, ancient religion or modern science. Neither view is satisfying. Perhaps it is the superb irrelevance of nature to the problems that obsess humanity that makes it such a sovereign remedy for frayed nerves and tired minds.

My life was in some ways very satisfying; much more so than many that civilization had to offer. But I felt that I had come rather too near the edge of society, and that I was prepared to sacrifice some freedom in return for the intellectual stimulus of a more civilized life. A very pedestrian conclusion, one must admit. Perhaps most people are wise to accept their values off the peg.

Many years later, sitting opposite an eminent ecclesiastic at a luncheon, I remarked that I must consider myself, like Churchill, a buttress rather than a pillar of the church. "Well," he said, smiling tolerantly, "It's cold outside." My thoughts went back to these years on the Peace River. Very satisfying they seemed in retrospect but it is quite true, it's cold outside. My conclusion was quite personal; Tom felt very much at home.

UNCHAGAH

Although the climax to my life on the Peace River was the realization that I preferred a more civilized existence, even at the cost of some of the freedom which I enjoyed, I was in no hurry to leave. I continued to enjoy my holiday, and, before events conspired to end it, I had an experience which, in that time and place, was as strange as any that could have happened to me. Far stranger and much more unlikely than being drowned in the river, killed by a grizzly, or adopted into an Indian tribe.

The life I was leading had sharpened my powers of observation, given me knowledge and expertise of a kind, and a great deal of low cunning, but it was devoid of any formal thinking. Analyzing one's feelings about nature can hardly be described as thinking in that sense of the word, and, after three and a half years of this kind of living I began to feel that my brain was rusting away. It happened that Tom, at some stage in his career, had studied civil engineering, and there were a number of text books on that subject in the cabin. I decided that, during my fourth winter, I would give myself a course in this subject as a mental exercise. In the event I exercised my mind in a very different fashion.

Early in the year a mining engineer named Guenther had come into the country to wash gold on Branham's Flat. Guenther was in his fifties, a dark complexioned man, silent and somewhat aloof. In his youth he had been an agricultural expert in the Cameroons, before they were lost to Germany; but an attack of blackwater fever[50] had driven him to seek a more northerly latitude, and he had spent many years in Alaska and Canada before I met him. He had heard about Branham's Flat, and had given up his job with the Winnipeg Electric Company to try his luck there.

It seemed rather pathetic to us who were living on the spot that anyone should have made this journey, for we knew it to be useless. There was gold on the Flat, but it was not a paying proposition to work it. So Guenther found it to be. He decided, however, to spend the winter in the bush, partly because it was very cheap living, partly in order to be first in the field in the spring. Jim offered him the old cabin where Tom and I had spent our first winter, and there he took up his quarters.

Although we were such near neighbours I should have seen little enough of Guenther in the course of a normal winter's trapping, but, early in the Fall, I discovered something about him that intrigued me. He practised astrology. Indeed, he had his whole life blue-printed, and could look up the planetary influences that were operating in any month or week as a man might look up a train in a timetable.

It is not surprising that some people consult astrologers, clairvoyants, and seers of various kinds, and "believe" in them without knowing anything about the theory or practice of their arts, but that an engineer should himself be an astrologer, and order his whole life according to his own forecasts, seemed to me astounding. Inevitably, I argued with him. Equally inevitably, I discovered something that I have often wished could be revealed to performers on television—that I did not know enough about the subject to discuss it intelligently. I asked Guenther what books I should read, sent to England for them, and soon became immersed in Ptolemy's Tetrabiblos and kindred works, almost to the exclusion of the winter's trapping.

In method, astrology is a science, and to concentrate one's mind on mastering a science which one believes to be nonsense is difficult, and would, I think, have been impossible for me in any other environment. But I was interested for two reasons. One was that, when one considers that, until a few hundred years ago, major decisions in civilized countries were commonly based on astrology, and that great scientists like Newton were more interested in astrology than astronomy, the subject has, at least, an historical interest. The other reason was that there might be something in it that modern scientists, busy about other and more rewarding studies, had missed.

One must also consider that the apparent validity of scientific hypotheses does not mean that they are true; indeed, since man's perceptive faculties and his knowledge lack perfection, perhaps they never can be wholly true. And, sometimes, they are wrong for the right reasons. For many hundreds of years it was thought that the earth was the centre of the solar system. This view did not persist because of the obscurantism of the church, but for what seemed the soundest of scientific reasons; for instance, there was no observable parallactic motion[51] in the stars. Even in the sixteenth century this reason was held to be valid by Tycho Brahe, said, I believe, to be the most acute observer of his own or any other time. The vast distances involved were not appreciated, and it was not, in fact, until the nineteenth century that parallactic motion was observed.

But Aristarchus of Samos, in the third century B.C., held that the sun was the centre of the solar system for what must have seemed to his fellow philosophers a very superstitious reason. The apparent motion of the planets, if real, could only be explained in terms of a very complicated system of cycles and epicycles. Aristarchus held that it was inconsistent with the perfection of the deity to have created such a system when the same effect could have been produced by arranging for the earth and planets to move round the sun in perfect circles.

As compared with the scientists, Aristarchus was right, although he did not have the last laugh. The planets do not move in perfect circles.

The great astrologers of antiquity were the Babylonians. They did not speculate about the geometry of the solar system—it is irrelevant to the astrologer, in any case—but they had observed the heavens for many hundreds of years and carefully noted their observations. They had done this, primarily, in order to construct a calendar, one of the first necessities of any civilization. They had not the means to observe any particular phenomenon exactly, but they had collated their observations over such a long period of time that they had achieved a great degree of accuracy. Their measurement of the lunar month, for instance, differed from that of modern astronomers by only one third of a second of time. These same men were astrologers, and it seems reasonable to assume that they approached that science with the same mental attitude, and one could not dismiss out of hand the possibility that they had achieved a similar result. As astrologers they did not say that a particular planet in a particular position in a horoscope "caused" certain kinds of events. They merely noted the correspondence and left it at that. It might be coincidence; it might be, as modern science sometimes suggests, that a long train of causation was involved.

By the beginning of the Christian era Babylonian astrology had received accretions both from the Egyptians and from the Greeks, and the whole of science had become a hotchpotch of theories, many of them based on far-fetched analogies, not on observation. It was in the second century A.D. that the great astronomer and geographer Claudius Ptolemy made the effort to sort the wheat from the chaff, which is outlined in his Tetrabiblos.

I was surprised and impressed by the Tetrabiblos. In the first place Ptolomy postulated that the superior influence would overcome the inferior, from which I deduced that if, say, a country went to war, many men would die violently in whose personal horoscopes this was not indicated. As to individuals, the major influence, he said, was heredity: "Under the same disposition of the Ambient and of the Horizon, each various kind of seed prevails in determining the distinct formation of its own proper species; thus man is born, or the horse is foaled." The second place he gave to environment: "Considerable variations are caused in all creatures by the respective places where they may be brought forth" and "Different modes of nurture, and the variety of ranks, manners and customs contribute to render the course of life on one individual different from that of another." No wonder he concluded that "It must not be imagined that all things happen to mankind as though every individual circumstance were ordained by divine decree," and in Centiloquoy: "It is not possible that particular forms of events should be declared by any person, however scientific … They only who are inspired by the deity can predict particulars." He confessed that many astrologers were frauds, and that all must often make mistakes owing to the complexity of the science. He advocated its study, not so that a man might resign himself to the inevitable, but so that, knowing the influences that were about to operate, he could make the most of the good, and avoid, as far as possible, the bad. All this seemed very sensible.

Guenther, at any rate, had no intention of abandoning himself to fate. He studied the stars as a pilot might study the weather, so that he might carry out his plans more easily and effectively. But he was a man of strong will, and obstinate; and his method of achieving harmony with the stellar influences was not as sensible as his desire to do so.

There is a branch of astrology known as "Horary Astrology" by which one can discover whether a certain course of action is likely to be advantageous or not, a piece of news true or false, and so forth. Guenther practised this, but I noticed that, whenever the stars differed from his preconceived opinion, he would worry about it for a day or two, and then sit down and make out another figure. This was quite wrong, and I was not slow to point it out. Guenther agreed, but he could not tolerate any difference between himself and the stars, and, at bottom, he put his own judgement first, and carried on until the stars agreed, at least on paper. I was reminded of the name given to the local astrologer in the village where I was born. He was called The Star-Maister.

It was on one of these occasions, when Guenther had consulted the stars several times on an expedition he proposed to make the following spring, that he came to me with a proposition. I was to provide the boat and outfit, and together we were to investigate a tributary of the Finlay River. I had heard of this place from old Louis Petersen. He had told me that it took almost the whole season to get there and back, that the gold was under heavy gravel, and needed machinery to work it. The only possibility was to take machinery in by air, and it was doubtful if this would be a paying proposition. I had no intention of making the trip, but I did not tell Guenther so. Instead, I glanced at my watch, and proceeded to consult the stars. The figure, fortunately, was most adverse; in particular, the position of Saturn was a most unpropitious augury for the financial results of the proposed partnership. I said nothing, but passed it over to Guenther. He looked at it for some time in silence, and then said; "Not too good, it is? Ah well, we must try again."

It was in this manner I spent my last winter in Canada, making only three or four trips over the trap-line. It was a weird business. After working until the early hours of the morning I would sometimes stroll out into the cold night air, and look up at the Moon and Venus blazing overhead, wondering how the idea that they influenced mundane affairs ever entered the head of man. Yet the Moon influenced the tides, and although we could describe it, and had given a name to it, the influence remained a mystery. Then I would return to the cabin, tune in the radio to a distant station, and continue to wonder.

The result of my studies was to qualify me to practise astrology if I wished, but there was not much I could do in the bush. I did, however, formulate a test of its validity so simple that I wonder why, in these days of newspaper astrology, I have never seen it suggested. The claim is that, if the exact time and place of birth is known, the position of the heavenly bodies will describe in a kind of celestial shorthand the influences in operation, and the times when they will be particularly active. It follows that, given the date and place of birth, and a man's biography, a competent astrologer should be able to work out the time at which he must have been born. There can only be one horoscope that can describe the man, and his career, and time the events in his life.

It would, no doubt, be hard to do this in the case of many men, but, if the man has had an outstanding career, it should not be difficult. Take, for instance, the case of the late Sir Winston Churchill. His biography is known; nothing more can happen to him under the sun. The day and place of his birth are known; the time, as far I know, is not publicly known. There should be no disagreement among competent astrologers as to when he must have been born. If that should agree with the record, (presuming there is one), allowing a margin of error in noting the time, say, of half an hour on either side, and if this kind of test can be successfully repeated, there is no need to indulge in abstruse calculations involving thousands of cases and the law of chance.

There is an Act of Parliament, passed in 1597, which reads that "All idle persons pretending that they can tell destynyes, fortunes or such other like fantasticall imanynacions shall be taken, adjudged and deemed rogues, vagabonds and sturdy beggars," and this was confirmed, by an Act in 1824. However, in 1936 it was held in the Lord Chief Justice's Court that this does not apply to forecasting in the popular press, since the popular press was not in existence in the reign of the first Elizabeth. Perhaps an Act should be passed requiring all who practise astrology for personal gain to undergo the test I have described on pain of being thrown into jail for false pretences if they fail.

Tom, on the occasions on which we were together, viewed my proceedings with amused tolerance. He thought, perhaps, that of all the ways of going mad in the bush, I had chosen one of the most harmless. The idea of going mad did not worry me; if one goes mad, someone else had to do the worrying. In fact, I came through this experience with nothing worse than the eyestrain caused by the constant use of logarithmic tables by candlelight, and, when I returned to civilization, a penchant for pestering my friends for information of a personal nature in order to check some astrological theory.

Early in May the mail carrier made his first trip by water, and Guenther, who had paid a visit to Hudson's Hope to get a grubstake, travelled with him. It was almost midnight when they arrived at Twelve Mile, and they made camp on the river bank, instead of coming up to the cabin. I had heard the engine, and strolled down to collect our mail, and Guenther and I sat on a log until the early hours of the morning, discussing astrology and prospecting. I gave him a Power of Attorney to stake a claim for me in case he found anything interesting, and few hours later he departed. He spent the spring prospecting up the Wicked River with a couple of tin plates slung across his shoulders to warn the grizzly bears of his presence, and a .44 Colt in case of an argument. It is bad grizzly country, but they did not get him. Later, he went up the Finlay, and for some time I heard no more of him.

By this mail I received news which necessitated my return to England for an indefinite period. It was a family matter, an obligation which I could not escape, and I did not leave unwillingly; but I delayed my departure as long as I could. It was a beautiful summer and autumn. Tom, who had spent his leisure during the winter hauling logs to build a house some way down stream near the river bank, was occupied in digging a cellar, and, under the cellar, a well, so that he could pump water into a tank in the roof, and have a tap, and a sink. Jim was having built a most palatial log house, forty-two feet square—the largest usable logs obtainable—and we worked on this during the intervals of working on the farm and fishing.

Building the New House at Twenty Mile, ca.1932.

Tom's house at Twelve Mile Flat.

Conditions on the Peace River were changing. Times were so bad outside==this was nineteen thirty two—that many men came into the bush to wash gold, and it was no longer safe to leave one's personal belongings on the river bank, or even to leave one's cabin unlocked. This change in social atmosphere made it easier for me to contemplate leaving, but I still delayed my departure. I had a longing to make one last trip up the river to Prince George.

It was in the Fall on this year that M. Bedeaux made a second big game hunting trip into the country, camping for a couple of days at Twenty Mile. The following year he brought in half a dozen Citroen trucks with caterpillar tracks of a type which had made a very successful journey across China, and had almost succeeded in crossing the Karakoram mountains into India. Unfortunately, they proved useless in the Rockies. M. Bedeaux bought a lot of land far in the north in a most inaccessible place. I suspected that he was preparing a hide-out for the next war. He was a millionaire and could indulge his fancies, which were not without reason.

As time went on it became evident that my only hope of making one last trip by river to Prince George lay in persuading Van Dyk, when he came through on his annual inspection, to turn round and go back up the river to Price George instead of continuing his journey by rail via Edmonton. I knew Van Dyk would be willing enough, unless he had business further east; he loved life in the bush; but I was doubtful whether or not the river would freeze over before we had time to make the trip.

Van Dyk arrived later than usual, and then had to go on to Hudson's Hope. It was late in October before we could make a start. The river was usually frozen by the fifth of November, but this year the weather remained exceptionally mild, and, although shore ice had formed, the river was still navigable. A few inches of snow had fallen, but the thermometer was still above zero. We decided to risk it, and, after saying good-bye to Tom, Jim, and his family, Van Dyk and I pushed off from the point at Twenty Mile.

I had come into this country in spring, and it seemed appropriate that I should be leaving on a dull October afternoon. It was the first time in years that I had travelled on the river when I was not running the engine myself, and I already felt a ghostly isolation from the life that I had been living. I took nothing out with me except my sleeping robe and a clasp knife. I remembered that I had waiting for me at Prince George the same suit of clothes that I had worn when I first arrived there, and I thought of the words in Ecclesiastes: "naked shall he return to go as he came, and shall take nothing of his labour, which he may carry away in his hand."

We reached the Ottertail River late in the afternoon and spent the night in Fred Chapman's cabin. Fred was not at home. Our first task next morning was to get the boat up the Ne Parle Pas rapid. This rapid gets worse as the water level falls, and at this time of the year it was impossible to run it, but, with Van Dyk at the engine and myself at the end of a hundred yards of line, we worked the boat up without mishap. A few miles upstream we passed the old camp site at Point Creek, and I noticed that the small cache which Tom and I had built when we first camped there was still standing. We did not stop to catch Arctic trout, but ran steadily all day and reached the Wicked River at dusk. We spent the night in the cabin there, and I walked round with a candle and reread all the scribbling on the walls which I had read four and a half years ago.

Ten miles of easy going in the early morning, and then the Finlay Rapid. Here again, I had to take a line out before we could get the boat through the rough water. It was only on occasions like these that I came to life. All day I lay in the boat, looking out over the water at the spruce clad mountains. I had no need of the knowledge of the river which I had acquired, and I felt that I should never need it again. It was the end of a chapter.

We passed Finlay Forks, and ran on up the Finlay to spend the night with Alan Mckinnon. Here, I enquired about Guenther, and found that he was camped on the south bank of the Finlay, undecided whether to come out by water or to cross the Wolverine Mountains on foot to Manson Creek. No one knew why he delayed so long in making up his mind, but, knowing Guenther, I had my own idea about that. I knew he did not intend to come back down the Peace River; he had told me so. I felt sure that he had decided to go to Manson Creek, but, on consulting the stars, had found them unpropitious. I had no doubt that he would remain in camp until the stars agreed with his pre-determined plan, regardless of the fact that any day the river might freeze over, or a heavy fall of snow make the mountain passes impassable.

The next day we turned back, ran down to Finlay Forks, and then up the Parsnip River. The first night we stayed with a Russian trapper; the second night we camped out. We had no tent, but we made a framework of poles and spread the boat tarpaulin over it, making a roof, back and sides. In front we made a fire about eight feet long, and after our evening meal, we lay silent in our sleeping bags, looking out over the embers at the snow-covered country, the dark spruce, and the stars. This was the last time I made camp in the bush, my last impression of the loneliness and peace of this north country. In the warmth of the sleeping bag I remembered with a shiver the discomfort of most English bedrooms at this time of the year.

The third night on the Parsnip we stayed with a trapper called Scott—an engineer who had come into the country many years before and had marked himself out a trap-line when men trapped more or less where they pleased. Unlike most of them, Scott had nursed his trap-line, and had built up a reserve of beaver, defending it from white man and Indian alike with his own right arm. He had been suspected of going to the extreme in this matter, but nothing had been proved.

Scott was tough, even by the standards of a country which does not encourage weaklings. He took a partner from time to time, but he was trapping alone the following winter when he broke a leg. Fortunately, the accident happened at his headquarters where he had plenty of food and firewood. He set the leg himself, and when he came out in the spring, the doctors told him that he had made a good job of it.

During the evening Scott asked me if I had ever shot caribou, and when I said that I had not—there being no caribou near the Peace River—he invited me to join him in hunting caribou the following year. I explained that I was on my way to England and that I did not expect to return. He looked at me steadily for a while. "Ah," he said, "You'll come back. I went out to the States few years ago, but I soon came back. Riding round in their goddamn cars, and going to their goddamn parties, that's all they know there. You'll come back." I did not dispute his criticism, but I knew that I should not return.

The following afternoon we left the Parsnip and ran up the Pack River to Fort McLeod. Snow was falling when we arrived, and, after covering the boat and engine, we went up to the Hudson's Bay Company's store. We spent the evening with the trader there and the night in the bunk house. Snow fell heavily all night. Van Dyk had to talk to the Indians, and, as it continued to snow all the following day, we stayed over until the next morning, and then crossed the lake and entered the Crooked River. We reached Davie Lake in the evening, and spent the night in a very draughty cabin on an island. Here we found two prospectors on their way out of the bush. They were cooking moose meat in a petrol tin, and, after, Van Dyk had satisfied himself that the proceedings were in order, we added some canned tomatoes, and joined them in eating the stew.

We started next morning as soon as it was light, and, after the few miles of dead water, began to run the riffles of the Crooked River. The water was low and freezing, but the boat was light. Van Dyk ran the engine as long as possible; when he had to switch off I jumped into the river and manhandled the boat, and Van Dyk poled. We stopped only once to make tea, and arrived, wet through and freezing, at Summit Lake just before darkness fell. As I stood steaming before a red-hot stove that evening I looked forward with a momentary feeling of satisfaction to the fleshpots of civilization. I slept that night in damp underwear for the last time. The next day we were in Prince George.

In Prince George I said good-bye to my friends there and took steps to cancel the Power of Attorney that I had left with Guenther. Then I returned to England, wearing the same clothes and carrying the same suitcase which I had brought with me when I came. The depression was in full swing. As the train wound its way slowly through the Rockies, I could see prospectors washing gravel in the freezing streams. There were only half a dozen first class passengers on the boat on which I crossed the Atlantic. I had my choice of innumerable cabins. Civilization wore a forbidding aspect.

For some years after I returned to England I took a small local paper called the Peace River Block News, and it was in the pages of this journal that I read the end of Guenther's story.

The account ran as follows:

> Mining Engineer's Body Found West of Hudson's Hope
> February 13th 1933
>
> *Word has just been received from Finlay Forks that the body of Mr Guenther, who was reported missing, has been found. Mr Guenther who hails from Winnipeg, came to Hudson's Hope about three years ago and was engaged in numerous mining projects throughout the district. Last November he set out from Finlay Fork apparently intending to strike out for Manson Creek to investigate some rumors of a gold strike there. Possibly thinking that by cutting down the weight of his pack he could make a quick trip and left the Forks with an insufficient food supply. The country he had to traverse is very rugged, for to reach Manson Creek it is necessary to cross the Wolverine range of mountains. No further word was received about him until some Manson Indians came to the Forks and reported that they had seen a white man dead in the snow.*
>
> *Vic Williams, the game guardian stationed at that point, accompanied by Iver Johnson and Indian guides, set out with a dog team to investigate. They were able to follow the faint imprints of the snowshoe tracks and found the body buried in snow, eighteen miles from Germanson Landing. He had evidently been following the old Moody trail and running out of supplies had become exhausted and lost his way wandering off on a pack trail, some three miles distant from the main route. The body was found by the remains of a small camp fire and it is supposed that becoming tired he built a fire and then drowsed off into the last long sleep. The North had claimed another victim. An idea of the country where he lost his life may be gained from the fact that it took Mr. William's party twenty three days to make the journey there and back a total distance of*

approximately 112 miles. It seems a hard fate that he should have passed away when he had so nearly reached his objective, but there is a limit to any man's endurance.

Guenther had been caught crossing the Wolverine Mountains in the heavy snow that had fallen while I was at Fort McCleod. He had spent many years in the bush, and no man of his experience would have attempted to make the trip at that time of the year without very good reason. It was only the previous year that two inexperienced men had perished in the Pine Pass in similar circumstances. I thought I knew the reason, and I tried, without success, to verify my suspicions. At the time I was corresponding with the then Commissioner of Police for British Columbia, the late T W S Parsons, and I wrote to him asking him if he could possibly send me Guenther's notebooks, but they had passed out of the possession of the police, and he could not do it. I have no doubt that, if I had been able to see them, I should have found all the astrological figures there which I felt sure Guenther had drawn while in camp on the Finlay.

If, indeed, this tragedy resulted from the curious division in Guenther's mind between his determination to go his own way, and his equally firm resolve to obtain the approval of the stars, it must be one of the most peculiar of modern times. If he had made the trip earlier in the season, he would have avoided the snow; if he had obeyed the stars in the first place, presuming that they were unpropitious, he would not have met with disaster. Guenther himself alone knew, and I have no doubt his knowledge confirmed his faith in the stars. I have often wished that I had taken a copy of his horoscope.

From time to time I had news of others I have mentioned in these pages. A few years after I left Canada, Don McDonald gave up trapping and migrated to the lower Peace River. Jack Pennington also ceased to trap, and settled at Hudson's Hope. Fred Cassie, who had trapped with Jack, took over the line at Crying Girl Prairie, and was drowned crossing the South Fork on ice. Charlie Jones and his wife moved to Vancouver Island where, despite his cigarette smoking, he lived to be over eighty. Jim Beattie continued to run his farm and, with the aid of a partner, his trap-line. He led his usual strenuous life until the spring of nineteen fifty when he wore out his heart and died at Hudson's Hope. Most of the children are married. Fred Chapman settled at Hudson's Hope, and we exchanged Christmas cards until the Christmas of 1964 when he sent me a photograph of himself, in close proximity to a bottle of rum, taken on his eightieth birthday. He looked hale and hearty, but there were no more cards.

Tom survived me as a trapper for ten years; then, he, too, left that part of Canada. It was in early spring that he took his farewell, and his mood was in keeping with the season. "I left my cabin at daylight," he wrote, "and climbed round the canyon of Eleven Mile creek. The sun rose just as I reached the highest point where I could look straight down some five hundred feet into the creek below, and across to the snow-clad mountains and timbered valleys beyond. The grip of frost had silenced somewhat the noise of the water, and the song of a dipper floated up to me out of the depths of the canyon. The same bird we used to see at Buckden[52], only grey instead of black and white. It stays here all through the winter wherever the current or a spring keeps any open water. Its song is rather like a weak, thin edition of a thrush's; not much in itself, but it blends with the sound of running water perfectly. They sing endlessly all through the winter and are a great favourite of mine. They seem to defy all natural laws, walking under water as though made of lead and the next moment swimming like a cork. In and out of the water they go in weather so cold you'd think they would freeze solid the moment they emerged.

"It really was a lovely day . . . I went up the creek seeing fresh tracks of marten, fisher, mink, and wolf. At noon I stopped, chopped a hole through the ice, and in ten minutes had caught a dozen ten-inch trout with a couple of feet of line and an imitation minnow. High up in the air two ravens soared giving at intervals their metallic mating call, and I saw several crossbills, rosy red like miniature parrots, and a few grouse. I had to tear myself away." He came out by plane.

Tom also left the Peace River to discharge a family obligation which he could not well avoid, and he left reluctantly. He never shared my feeling about nature, or, at worst, he still preferred it to human society. To him a human being was always a potentially more dangerous and incalculable animal than a grizzly bear, and much more likely to interfere with him. That is true enough, as far as it goes. It went far enough for Tom.

But I have never seen any reason to alter my judgement, although civilization becomes more oppressive year by year, and seems likely to continue to do so. Increasing mechanization has obscured the ends of life, and unendurably complicated the means. This, and our gradual evolution towards the totalitarian state, seem, as I have already said, to have deprived man of some of his dignity and significance. But there is significance nowhere else in the universe. In nature there is a healing and sustaining power which is a powerful antidote to the tensions of the man-made world. But it only provides a rest-cure, and man must inevitably return to deal with those problems, which it is his business, and, fundamentally, his desire to solve.

In those September days at Hudson's Hope when I had pondered on the future of the country it had seemed to me that, rich in natural resources as it was, development lay far away, probably not in my lifetime. The wheat farmers to the east wanted an outlet to the Pacific, but as yet there were too few of them to justify the capital expenditure involved. A government sponsored scheme of settlement in the area was mooted, but this seemed likely to remain a pipe dream, and a rich mineral strike was a matter of pure chance. Sixty miles down the river was the small town of Fort St John, and the Northern Alberta Railway was slowly creeping from Pouce Coupe to Dawson Creek, about forty miles to the south, but there seemed no reason why it should cross the Rockies. A road on which tourists could drive north from Vancouver to Prince George and make a circular trip via the Peace River to Edmonton was also in the air, but even a direct connection between Edmonton and Prince George by road had not yet been made. Later, during the years of depression, development would have seemed to me still further away, if I had thought about the matter at all.

The possible effect of the one thing that drives man to heroic achievement, and for which money can always be found, had not occurred to me, and, if it had, I should have discounted its impact on the Peace River Block. I mean war. A rich gold strike would have made dramatic changes; what actually happened was like a chapter out of the Arabian Nights. Until 1938 the development of the Tourist traffic formed the background to most of the plans for roads in this country, but in that year when war seemed inevitable, the government of the United States began to worry about the isolated position of Alaska, and, in the following year, various routes for a road were considered. In nineteen forty Russia was reported to be establishing an air base on the Big Diomede Island in the Bering Straits, only a mile and a half from the Little Diomede Island, an American possession. The entry of Japan into the war a year later brought matters to a head.

On the ninth of March, 1942, with the temperature about 30° below zero, U.S.A. General Hodge arrived at Dawson Creek. By November 28th of that year a pioneer road, sixteen hundred and seventy-one miles in length, passable at ten mph. for an army truck had been constructed from Fort St John to Fairbanks in Alaska. The following year it was widened to twenty-six feet. This road crosses the Rockies amid mountain peaks ten thousand feet high, but its highest point is only four thousand two hundred and twelve feet, and its average height about two thousand feet above sea level.

Oil was discovered at Fort Norman near Great Bear Lake, and in view of the possible requirements of the American fleet, it was decided to refine it and pipe it to the Pacific coast. The waterways might have been used for the transport of materials, but this would have been too slow. So, a thousand miles of rough road was constructed from the town of Peace River to Fort Norman for the transportation of equipment, and the oil piped five hundred and fifty miles to White Horse in Alaska on the other side of the Rockies.

UNCHAGAH

There are two hundred bridges on the Alaska Highway. The first of them crosses the Peace River at Fort St John. Later, it was swept away by the river and had to be rebuilt. Great changes took place in the area affected. At the beginning of 1942 the population of Dawson Creek was about four hundred; there were twenty thousand people there by the end of the year. At Fort Nelson, a mere post in the bush, there was a similar influx. But, when the effort had been made, this immense spate of activity ebbed away. For some years after the war the oil strikes in the Turner valley south of Edmonton put Norman Wells and the Athabasca tar sands in the shade. The Alaska Highway was open for tourist traffic during the summer to those whose cars were certified fit to make the trip.

Gradually, however, development was resumed, and the discovery of the immense deposits of natural gas in the vicinity of Fort St John assured the future of the country. Dawson Creek, which is just inside the boundary of British Columbia, was already connected by road and rail with Edmonton, Alberta, and by road with Fairbanks, Alaska. In course of time the old dream was realized, and the connection made with Vancouver. A road was driven north from Prince George to Fort McLeod, and then east through the Pine Pass to Dawson Creek. Later, even the Pacific Great Eastern Railway came to life, and followed much the same route. In 1958 the wheat farmer in the Peace River Block was at last able to ship his wheat to Vancouver by a direct route, but the dream of Captain Butler, Louis Petersen, General Sutton, and so many others was not to be realized; there is no railroad through the Peace Pass.

I read about this, and saw in photographs the derricks, refineries and grain elevators that dotted the land. It did not affect me emotionally, for to me prairies have no appeal. A herd of buffalo might have seemed more romantic, but a drilling rig did not offend me. Nothing happened to spoil my country west of Hudson's Hope.

Nothing, as yet. And now that the railway was running through the Pine Pass, it seemed that the upper Peace River might escape the attention of civilized man. But this was not to be. I had noted the potential for generating water power at the Peace River Canyon, but I had no conception of the magnitude of the scheme that was eventually drawn up. In October 1957 the government of British Columbia signed an agreement with the Wenner-Gren B.C. Development Company commissioning it to carry out a survey so that before the end of 1959 firm contracts could be entered into for the purpose of damming the Peace River, and turning it, and many miles of the Parsnip and Finlay rivers, into one vast lake.

The W.A.C. Bennett Dam and Williston Lake
- Aerial Photo by Del Herbison – Hudson's Hope, B.C.

To dam the Rocky Mountain Canyon for the purpose of generating electricity for the Peace River Block seemed a likely development once the growth of population justified it. To re-create the ancient barrier of the Rockies to the waters flowing from the west, flooding the valleys of the Peace, the Parsnip as far as the Pack River, and the Finlay for a similar distance, was beyond my imagination. It would make a lake some two hundred and sixty miles long, and generate twice as much electricity as the Grand Coulee, the largest dam in the United States. In due course the British Columbia Hydro and Power Authority took over from Wenner-Gren, and this is what has been done. By now the Peace River Trench should be under six hundred feet of water. Children, are now, I suppose, dead, and the country where they lived has vanished too. Twelve Mile, Twenty Mile, Carbon River, the Ne Parle Pas, and the Finlay Rapids, all are now submerged. Although Tom stayed in Canada, neither he nor I ever went back to visit them, nor do I think that we should ever have made that sentimental journey. There is seldom any point in going back. It only over-lays a happy memory with one, perhaps, less happy. Now we cannot go back if we would. That is all the difference it makes, and I am glad it has happened so. Derricks, and the refuse of mining camps, would have been a desecration. Now, only holiday-makers will sail over the water, while beneath, like part of the lost kingdom of Lyonesse[53], lies the land that we knew.

"And one remembers and forgets,
But 't is not found again,
Not though they hale in crimsoned nets
The sunset from the main."

The End

Appendices

Appendix A: An Account of the Origin of the name Peace River 136

Appendix B: A recollection of Indian Lake by, Tom Stott, ca.1951 138

Appendix C: Al D. Young – Prince George Taxi Driver. 144

Appendix D: Last photograph and inscription on Christmas card, ca. 1950 148

Appendix A

An Account of the Origin of the name Peace River

AN EXTRACT FROM a speech to the Empire Club in Toronto by the Right Reverend E. F. Robins, 21 February, 1929:

> *"When the Indian roamed over the pathless ways of the prairies and crossed rivers and lakes and mountains there was a conflict so severe that it remained long in their memories. The southern men pressed the northern men back and back until it was impossible for the northern men to retire any further. At a certain point on the Peace River there is an immense outcropping of gypsum. I passed up that river in 1912, and it took me three months away from civilization or contact with the outside world. I was very much impressed when I saw a great outcropping of gypsum on the southern banks of the Peace River for the Peace is flowing west to east at that point. The northern men made their last stand because they could not cross the river. They gave in and made a pact that from that time a truce should be observed, and never be broken; and in the flowing, beautiful imagery of speech of which the Indian is a master he made his legal promise that this great white rock of gypsum which stands by the flowing waters which here pass on to the distance where the sun rises in the great space beyond --- that this great white rock and the flowing waters should ever bear record to the fact that the Peace was made, and that the waters that flow by the white rock should be called the Waters of the Peace. That is the Indian origin of the Peace River."*

Some recorded Indigenous words which translate into Peace or Peace River:

"Unjigah or Peace R^r." shown on a map in Alexander McKenzie's book of 1801 "Voyages from Montreal", and reproduced in Derek Hayes' "British Columbia - A New Historical Atlas", (2012).

Unchaga (Cree for "Peace"): from an article in The Canadian Encyclopedia by David W. Leonard (2008).

Unchaga, Unchagah or Unjigah may have been used by the Dane-zaa, Dunne-za (Beaver)[54] as noted in several pieces by historian Dorthea Horton Calverley (1903 – 1989). The pieces, including one titled 'Unchagah', were obtained from The South Peace Historical Society. The Dane-zaa territory is mostly around the Peace River east of the Rocky Mountains.

Unjika or Unjiga may have been the Tse'khene (Sikani) for Peace River in their territory west of the Rocky Mountains including Fort McLeod, Fort Graham and Fort Ware. From a PhD Thesis titled "Language, Legends, and Lore of the Carrier Indians" submitted by J.B. Munro, Ottawa, University of Ottawa, 1944.

The validity of the Dane-zaa and Tse'khene names noted above is uncertain and difficult verify due to the passage of time and the relatively small number of native speakers remaining in the region. (Ref. Pg. 22)

Appendix B

A recollection of Indian Lake by, Tom Stott, ca.1951

Indian Lake

I suspect that many anglers, in these overcrowded times, cherish the hope that, one day, they may discover some lost lake or lonely stretch of river where fishing conditions still prevail similar to those that make us sigh with envy when reading the stories of Frank Forester and other writers of a bygone era. For some years I shared the privileges of such a water with none but the mink and otter, the loon and the osprey; as the family of Beaver Indians who had made it their headquarters, having had a run of bad luck, had deserted the place as ill-omened, some time before I took it over as part of my trapping territory.

Indian Lake[55], which now, since aerial survey, goes under another name, is one of those ice-cold mountain lakes with no definite inlet, fed by springs and several small streams, one of which is heavy lime which may account for the fertility of the water. Lying due East and West, it is about three miles long by five or six hundred yards wide at the blunt eastern end, becoming gradually narrower and shallower towards the West, where the outlet is blocked by a series of beaver dams: a place much haunted by moose. At the eastern end is shelving beach of soft, broken-up shale; elsewhere the shores are steep and rocky, the water often being ten feet deep a couple of yards from shore. About five miles below the lake, Indian Creek goes over a fifty foot fall. Below, the fish are the usual dolly varden, rainbow trout and arctic grayling of the district; but in the lake lives a variety of rainbow trout whose flesh, probably due to something in its diet, is as red as that of a sockeye salmon. There are also lake trout and suckers for them to feed on, although the latter are so little in evidence that I did not note their presence for over a year.

For the most part these rainbow are as sporting a fish as ever took a fly. At the prick of a hook they go skittering across the surface at a terrific speed; leaping, diving, and going through all the aquatic acrobatics imaginable. About one in ten, however, will show much less life and, on examination, will be found to have white or pale pink flesh; otherwise it will be indistinguishable from the rest – a parallel with the white spring salmon, which even an old fisherman cannot be sure of telling from a red without nicking the flesh. They are not large fish these rainbows, averaging about three-quarters of a pound, but as game or pan fish are unbeatable and I kept a fly-rod and tackle in my cache at the lake on their account. It seemed an outrage to take them on coarse tackle like the lake trout, which I caught up to twenty-five pounds in weight in the deepest part of the lake where the water is from a hundred to a hundred and forty feet deep. I had a dug-out canoe, made from the largest cottonwood log available, ten feet by two, in which I could keep right side up if I held my breath and made no sudden movements. By fishing on an out-rigger, south sea fashion, I made it stable enough to fish from and even sail on occasion.

The lake had at one time been a beaver paradise and there were three large houses at intervals along the northern shore. These were used in rotation by the resident family who, having eaten all the aspen and willow close to the lake and not liking alder, were forced to make long and dangerous trips up the steep hillside. It was whilst standing on one of these houses that I first noticed the suckers, for a large school was swimming slowly in and out of the net-work of submerged sticks protruding from the house. They formed, no doubt, the main food supply of the lake trout. I discovered, more or less by accident, that the smaller lake trout could be taken in shallow water on fine tackle late in the season and provided fair sport. I had arrived at my cabin, at the south-east corner of the lake, about an hour before dark one stormy day in September. I had counted on fish for supper but the prospects were grim as a strong westerly wind was sweeping down the lake, throwing great waves far up the shingle beach. There was no casting a fly in such weather so I put on a small spoon and, throwing across and partially into the wind, got it some twenty feet out. I allowed the light spoon to sink for a moment and then commenced to reel in. At the first movement there was a solid tug and, after a short struggle, I landed a three pound lake trout. I repeated the process thirteen times, hooking a fish at every cast and landing eight of them, all identical in size. How long I could have kept it up I don't know, as there seemed no end to them; but the gut, frayed by many sharp teeth, parted at last and, as it was quite dark, I thought it time to quit. I suppose they must have gathered to feed on schools of small fry that usually inhabited the shallows off the shingle beach but had been forced to seek deeper water in order to avoid being stranded by the waves. These lakers, particularly the large ones, were excellent eating, either fresh or smoked, although they were not as much sport to catch as the rainbow.

The surrounding country was the home of many game animals: moose, bear, deer and mountain goat, but there were no mountain sheep and caribou were very rare. Ruffled and blue grouse, spruce hens and ptarmigan were also found there and the lake always held a few ducks, although it was off the direct route of migration. Wild gooseberries and red currents grew near the shore whilst on the surrounding mountains there were blueberries in profusion. In the fall it was a camper's paradise.

I shall always remember the last evening I spent at Indian Lake. It was late August and had been a hot day. Sated, for once, with both fish and fishing I lay on the shingle beach watching, through my binoculars, the antics of a family of goats on the rocky face of Triangle Mountain[56]. The shale, soft enough to be crumbled in the fingers, gave easily to the body, providing a comfortable couch and was also a safe place for a fire to cook on, even in the driest weather. There was no wind, but the smooth surface of the lake was broken by the rising innumerable trout. A hundred feet up a fish hawk hovered. Three times he dived, sending up showers of spray, before he was, at last, successful and flew off, with a shining trout in his claws, to his supper table on the wind-shattered top of a spruce tree. Slowly the sun sank until only the white goats on the mountain top still enjoyed its rays. From far down the lake came the wild call of the loon, emphasizing with its clamour the absolute calm and peace. Fewer trout were rising now and I see the V-shaped wake of the first beaver out that evening, on an exploratory trip to see that all was well before venturing ashore to forage. Then, from somewhere far to the east, came the lonely howling of a wolf...

Nowadays, whenever the petty irritations and exasperations of modern civilized life drive me to the verge of madness, I recall this scene to recover some of that lost serenity. (Ref. Pg. 94)

FINIS

Tom Stott at Indian (Carbon) Lake with a large lake trout.

Rainbow trout and a small lake trout.

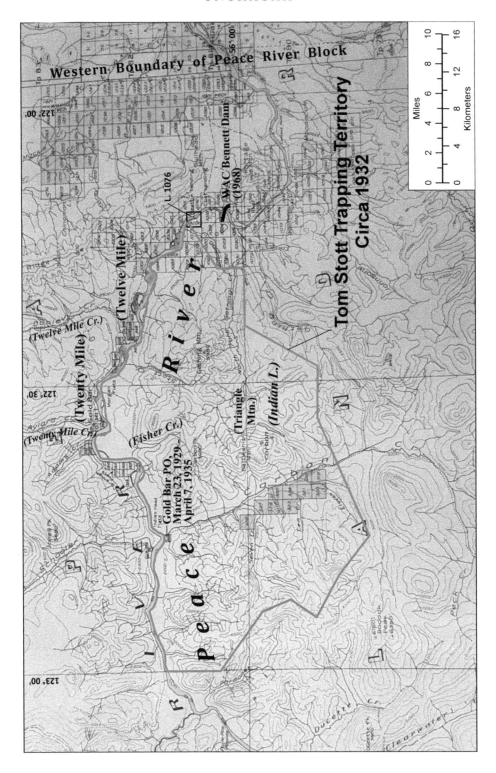

Tom Stott Trapping Territory as shown on portion of BC Peace River Map 3E, circa 1944; 22374C courtesy of the Royal BC Museum and Archives.

Appendix C

Al D. Young – Prince George Taxi Driver

ONE OF THE taxi drivers Henry Stott and his brother Tom met in Prince George in 1928 was likely Al D. Young. Al was well known in his time as a stagecoach driver, mostly on the Cariboo Road.

In 1930 Henry and Tom hired Al Young in Prince George to take them to Quesnel. From Quesnel they would take the train on PGE Railway to Squamish on their way to Vancouver.

In the B.C. Provincial Archives the voter's lists for both the provincial elections of 1924 and 1928 show Al Young's occupation as taxi driver. Of the half dozen taxi drivers recorded in the 1928 voters list for Prince George, only Al Young's name appears with his occupation as stagecoach driver in the voters list of 1920.

The following account is based on articles in The Vancouver Daily Province and from records in the B.C. Provincial Archives. The photographs and newspaper clipping are from copies in the Prince George Museum.

Al Young was born at a stage station in Colorado, U.S.A. in 1867, on the route between Kansas City, Missouri, and Santa Fe, New Mexico. The son of a stage driver he became a stage driver at age 18 and continued in that occupation for nearly 30 years. During his career he drove stages for nine different companies in five states and in B.C. Once the arrival of trains in the U.S. put an end to stage driving, Al came to Canada. He continued driving for several firms including the old B.X. Company. By the early 1920's the train and the automobile had put an end to stage driving in B.C. Al then took up taxi driving in Prince George.

A newspaper photograph shows Al holding a whip in his huge right hand. Another photograph shows Al freighting on the Cariboo Road, in which he is single-handedly driving four teams of horses pulling three wagons. In the 1920's and 1930's Al would take part in parades demonstrating his skills handling six-horse stages including those held by the Cariboo pioneers in 1926, and the 1936 Cariboo Gold Rush parade put on by the B.C. Chamber of Mines in Vancouver.

Al Young died in Lytton in 1947, age eighty, and is buried in Lillooet. This seems appropriate since Lillooet is Mile "0" of the old Cariboo Road to the goldfields of the B.C. interior. (Ref. Pg. 107).

*Al Young freighting on the Cariboo Road,
driving four teams of horses, pulling three wagons, ca. 1910.*

Al Young driving a stagecoach arriving at Quesnel, 1910.

ARMED with two four-horse whips, a pair of stage-driver's buckskin gloves and an old-time "treasure" bag such as carried on Cariboo stages, Al Young, veteran stage driver of many roads has arrived in Vancouver. He is here to drive the six-horse stage coach in the Cariboo Gold Rush parade arranged by the B. C. Chamber of Mines July 20 to 25.

AL. YOUNG

The coach, which has been lent by Lord Martin Cecil of the 100-Mile House Ranch, arrived here on Saturday night's P. G. E. train. It will form part of the old-time equipment which will feature the parade, with Al Smith driving six horses.

The Dufferin Coach, famous "de luxe" vehicle of B. X. stage days, built expressly for Lord and Lady Dufferin on the occasion of their visit to B. C. in 1876, was afterwards used by the stage company for its "hi-muck-a-muck" patrons—or those who desired for some reason or another, to give an appearance of affluence. They paid for this privilege twice the rate of ordinary vehicles.

It will be driven by Fred Tingley, son of the late Steve Tingley. Fred drove stage for his father on the Cariboo road for several years.

Al obtained the four-horse whips from Alphonse Hautier of Lytton and now has them completely remodelled. With these instruments Al is able to flick a fly off a night leader without touching a hair of the animal. But he can also make its finely platted silk "popper" felt if occasion demands.

One of Al's chief concerns, when he took on the job here, was whether or not he would be able to secure proper buckskin gloves. He was fortunate, however, in obtaining a pair in Ashcroft.

"I soaked 'em in water, put 'em on an' slep' in 'em and this mornin' they jest fit fine," Al said today.

Al will spend his spare time checking over and arranging the stage equipment which has been gathered from various parts of Cariboo and the interior.

Veteran Driver, With His Gloves and Whips, Will Be In Charge of Six-horse Stage In Gold Rush Parade.

Vancouver Province article and photo of Al Young as he prepared for a demonstration of stagecoach driving, ca. 1928.

Appendix D

Henry's last photo of Twenty Mile as he was leaving the Peace country in October 1932.

UNCHAGAH

Inscription on Christmas card, ca. 1950, from Henry Stott and his wife Olga with photo looking back at Twenty Mile:

> *"O mihi praeteritos referat si Jupiter annos."*
> (If only Jupiter would restore those bygone years to me.)
>
> From: Vergil, Aeneid VIII . 560
> Epic Poem of the Classical Roman Poet
> Based on Homer's Iliad
>
> Publius Vergilius Maro (70 – 19 B.C.)
> P. Vergil Maroni
> AENEIDOS LIBER OCTAVVS
> Ref. Pg.124

Index of Images

1. *Henry Stott, 1947* — x
2. *Tom Stott on SS Laurentic, 1928.* — 5
3. *Mt Selwyn as seen from the Wicked River.* — 27
4. *Wicked River falls.* — 30
5. *Looking west up the Peace River from Schooler Creek.* — 35
6. *On the Peace River in winter, looking west from Schooler Creek.* — 36
7. *Map of Water Route to the Arctic.* — 38
8. *Group at the old boat landing of Twenty Mile, 1931.* — 39
9. *Charlie and Madge Jones in front of their house at the mouth of the Carbon River.* — 41
10. *Shorty Kierce and Tom Stott at Charlie Jones' home.* — 41
11. *Toulie and Girlie Beattie at Charlie Jones' home, ca. 1928.* — 42
12. *Up at Carbon River – Photo from Hudson's Hope Museum.* — 43
13. *Trapline Registration Certificate, 1931* — 46
14. *Henry with Olive and Girlie Beattie at Twenty Mile.* — 48
15. *Twenty Mile, feeding stock.* — 49
16. *Tom up on the Clearwater River with rainbow trout.* — 51
17. *Tom's Trapline* — 52
18. *The 'Pines' Cabin. – about twelve miles from Twenty Mile.* — 68
19. *Donald McDonald at his headquarters cabin at Crying Girl Prairie.* — 70
20. *Tom landing his boat near his house at Twelve Mile.* — 71
21. *Tom holding antlers at front gate to winter home.* — 72
22. *Jim Beattie's Twelve Mile cabin and cache.* — 75
23. *Christmas at Twenty Mile, 1930* — 76
24. *Christmas at Twenty Mile, 1931* — 77
25. *Tom skinning fisher on South Fork divide — Graham River.* — 78
26. *Henry carrying pack at first cabin up Twenty Mile Creek.* — 81

27.	*Tom with large Lynx, February 1932.*	82
28.	*Furs on Cache at Twelve Mile winter home — martin, fisher and lynx.*	83
29.	*Map of some District Lots at Twelve Mile.*	96
30.	*B.C. Names Card — Stott Creek*	97
31.	*Henry with moose, ca.1930.*	114
32.	*Building the New House at Twenty Mile, ca.1932.*	122
33.	*Tom's house at Twelve Mile Flat.*	123
34.	*The W.A.C. Bennett Dam and Williston Lake.*	132
35.	*Tom Stott at Indian (Carbon) Lake with a large lake trout.*	141
36.	*Rainbow trout and a small lake trout at Indian Lake.*	142
37.	*Map of Tom Stott Trapline*	143
38.	*Al Young freighting on the Cariboo Road, ca.1910.*	145
39.	*Al Young driving a stagecoach, arriving at Quesnel, 1910.*	146
40.	*Vancouver Province article and photo of Al Young, ca. 1928.*	147
41.	*Henry's last photo of Twenty Mile as he was leaving the Peace River Country, October 1932.*	148
42.	*Henry Stott, 1959*	156

Endnotes

1. Unchagah: one of several similar First Nations words which translate into Peace or Peace River.

2. my brother Tom: Thomas Staynes Stott, b. 1903, Bradford, Yorkshire - d. 1971, Victoria, British Columbia.

3. gralloch: disembowel

4. Peace River Block: a block of land (3.5 million acres – 14,000 km^2) given to the Canadian Pacific Railway to compensate the CPR for non-arable land in the forty mile wide land grant received for building the transcontinental railway.

5. Captain Butler: Captain William F. Butler, in the 1870's he was commissioned to survey across the northwest of Canada.

6. The Upper Peace River: was generally the Peace River Country between Finlay Forks in the west and Hudson's Hope to the east.

7. Alekhine: Alexander Aleksandrovich Alekhine 1892 – 1946: Russian chess grandmaster and fourth World Chess Champion.

8. Grand Trunk Pacific Railway: now the Canadian National Railway (CNR).

9. the man: Al D. Young – former stagecoach driver.

10. Chilacotans: the Indigenous people of the Chilcotin or Tsilhqot'in First Nation.

11. Siwash: Chinook Jargon for an Indigenous man; used by early traders. It has been retained in the book because it was a word known to the author during his time in Canada. The family of the author mean no offense by retaining it. They accept that now it can be regarded as derogatory.

12. 150 proof spirit: 75% alcohol by volume.

13. 'squaw fish': Northern Pikeminnow

14. Izaak Walton: (1594 – 1683) English author of 'The Compleat Angler'.

15. John Jacob Astor: (1763 – 1848) prominent American businessman, merchant and fur trader.

16. Sir George Simpson: (1787 – 1860): the Canadian governor of the Hudson's Bay Company 1820 – 1860).

17. old Indian chief: Chief Dan Dick.

18. 'two Indian tribes': in pre European times the Cree and Dane-zaa (Beaver) First Nations made a peace at Peace Point on the Peace River north of Lake Claire and northwest of Fort Chipewyan on Lake Athabasca in what is now north eastern Alberta.

19. 'Regnier': Mathurin Regnier (1573 – 1613): French satiric poet.

20. Translation of the epitaph of Mathurin Regnier:

    > I have lived thoughtlessly,
    > Letting myself yield gently
    > To the good old natural law,
    > And thus greatly wonder why
    > Death should take notice of me
    > Who never took notice of her.

21. 'Punch': British weekly magazine of humour and satire, 1841 – 2002.

22. Twenty Mile Creek: Aylard Creek.

23. Carbon River: also known as Carbon Creek.

24. Twenty Mile: later known as Gold Bar when the Gold Bar Post Office at Charlie Jones' home at the mouth of the Carbon River was moved to Twenty Mile in 1935.

25. Branham's Flat: original name, Brenham Flats.

26. Rocky Mountain Canyon: also known as the Peace or Peace River Canyon, site of the W.A.C. Bennett and Peace Canyon Dams.

27. Cockermouth: a town in Cumbria in northwest England.

28. Twelve Mile Creek: Dunlevy Creek.

29. toxic goiter: Graves' disease; overactive thyroid gland.

30. Lord Rhondda: David Alfred Thomas, 1st Viscount Rhondda (1856 – 1918) Welsh industrialist.

31. Grenfell cloth: British manufactured high quality close woven cotton twill for outdoor clothing.

32. South Fork of the Halfway River: the Graham River.

33. Eskimo: Inuit; Inuktitut, "the people"

34. Bond Street: a street which is home to the most elegant and expensive shops in the West End of London.

35. valenki: traditional Russian winter felt boots.

36. "grumbling appendix": old term to describe chronic appendicitis.

37. eldest girl: Toulie (Louise) Hamilton, nee Beattie

38. queering Jim's pitch: originally to interfere or spoil the business of a tradesman or showman.

39. "Hugh the Drover": an opera written in 1914 by English composer Ralph Vaughn Williams.

40. Stornoway: a town on the Isle of Lewis, in the Outer Hebrides of Scotland.

41. "Mrs. Beeton": English journalist, editor and writer 1836 – 1865. Best known for her book "Mrs. Beeton's Book of Household Management".

42. whisky jacks: members of the Jay family, also Grey Jay or Canada Jay.

43. froe: a tool used for splitting shakes from blocks of wood.

44. the mountain: Battleship Mountain, (Triangle Mountain).

45. a lake: Carbon Lake, (Indian Lake).

46. ankylosis: abnormal adhesion and rigidity of a joint due to fusion of the bones.

47. peevee: a loggers tool for rolling logs.

48. Aimèe Semple McPherson: a media celebrity and evangelist of the 1920s and 1930s.

49. skeg: a sternward extension of a keel of boats with a rudder mounted on the centre line or the lowest point of an outboard motor or the drive of an inboard/outboard.

50. blackwater fever: a complication of malaria in which red blood cells burst into the bloodstream, releasing haemoglobin directly into the blood vessels and into the urine, frequently leading to kidney failure.

51. parallactic motion: the apparent motion of stars due to the orbital motion of the earth.

52. Buckden: an English village on the east bank of the River Wharfe in the Yorkshire Dales National Park.

53. Lyonesse: a country in Aurthurian legend said to border Cornwall and to have sunk under the water.

54. Dane-zaa (Beaver) and Tse'khene (Sikani): among the Northwest Canadian Athapaskan group of languages.

55. Indian Lake: a local name for Carbon Lake.

56. Triangle Mountain: a local name for Battleship Mountain.

Henry Stott, 1959

HENRY BROADBENT STOTT was born in 1895 in Bradford, West Yorkshire, England the eldest of four children. He was educated at Bradford Grammar School, where he was a brilliant scholar, and read History at Lincoln College, Oxford.

The First World War interrupted his studies when conscription was introduced in 1916 and he was imprisoned as a conscientious objector. After the war, although not a Quaker, he worked with the Friends War Victims Relief in Poland until 1923.

After finishing his degree he returned to Bradford until 1928 when his brother Tom, who had been farming in British Columbia, paid a visit home and persuaded Henry to join him in Canada.

Henry already had a passion for fishing but, as unlikely as it seems, it was in the bush country he was introduced to Astrology which he practiced with great skill until suffering a stroke in 1975. The last horoscope he prepared for himself predicted both the stroke and his death in 1981.

Unchagah is the story of the adventure he shared with his brother Tom in the Upper Peace River country of British Columbia.